BASIC MATHEMATICS FOR SCIENCE

Books by M Nelkon

Published by Heinemann
ADVANCED LEVEL PRACTICAL PHYSICS (*with J. Ogborn*)
SCHOLARSHIP PHYSICS
OPTICS, WAVES, SOUND (formerly *Light and Sound*)
MECHANICS AND PROPERTIES OF MATTER
PRINCIPLES OF ATOMIC PHYSICS AND ELECTRONICS
REVISION NOTES IN PHYSICS
  Book I. Mechanics, Electricity, Atomic Physics
  Book II. Optics, Waves, Sound, Heat, Properties of Matter
GRADED EXERCISES AND WORKED EXAMPLES IN PHYSICS (with Multiple Choice Questions)
NEW TEST PAPERS IN PHYSICS
REVISION BOOK IN ORDINARY LEVEL PHYSICS
ELEMENTARY PHYSICS, Book I and II (*with A. F. Abbott*)
MATHEMATICS OF PHYSICS (*with J. H. Avery*)
ELECTRONICS AND RADIO (*with H. I. Humphreys*)
SOLUTIONS TO ADVANCED LEVEL PHYSICS QUESTIONS
SOLUTIONS TO ORDINARY LEVEL PHYSICS QUESTIONS

Published by Hart-Davis
PRINCIPLES OF PHYSICS
EXERCISES IN ORDINARY LEVEL PHYSICS
C.S.E. PHYSICS
REVISING BASIC PHYSICS
SI UNITS: AN INTRODUCTION FOR ADVANCED LEVEL

Published by Edward Arnold
ELECTRICITY

Published by Blackie
HEAT

Basic Mathematics for Science

for Ordinary level, CSE, and technical students

M. NELKON M.Sc.(Lond.), F.Inst.P.

*Formerly Head of the Science Department
William Ellis School, London*

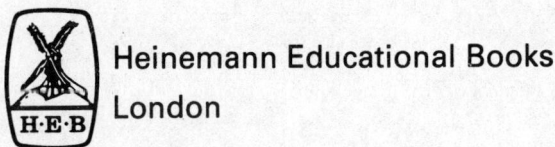

Heinemann Educational Books
London

Heinemann Educational Books Ltd

LONDON EDINBURGH MELBOURNE AUCKLAND TORONTO
HONG KONG SINGAPORE KUALA LUMPUR NEW DELHI
NAIROBI JOHANNESBURG LUSAKA IBADAN
KINGSTON

ISBN 0 435 50610 2
© M. Nelkon 1978
First published 1978

Published by Heinemann Educational Books Ltd
48 Charles Street, London W1X 8AH

Printed in Great Britain by
J. W. Arrowsmith Ltd, Bristol BS3 2NT

Contents

Page

Preface

ARITHMETIC

1 Fractions: adding, subtracting, multiplying, dividing   3

    Adding and subtracting fractions   4
    Multiplying fractions   5
    Dividing fractions   7

2 Decimals: adding, subtracting, multiplying, dividing   10

    Decimals and fractions   10
    Metric system and SI units   11
    Adding and subtracting decimals   12
    Significant figures   13
    Measurements and significant figures   14
    Multiplying decimals   14
    Order of accuracy   15
    Dividing decimals   18

3 Ratios and percentages. Chemistry calculations   21

    Ratios   21
    Ratio calculations   22
    Percentages   23
    Efficiency of machines   23

    Chemical calculations   24
    Percentage mass composition   24
    Law of multiple proportions   25
    Chemical formulae   27

4 Indices   29

    Powers of 10   29
    Multiplication with indices   30

CONTENTS

Brackets and indices	30
Division with indices	31
Rules for directed numbers	33
Negative indices	34
Fractional indices	38
Some areas and volumes	39
Areas	39
Volumes	40
Surface area to volume ratio in warm blooded animals	41

5 Logarithms — 43

Logarithm tables	44
Antilogs	44
Multiplication and division using logs	45
Logs of decimals or fractions	46
Multiplying decimals using logs	47
Dividing decimals using logs	48
Powers using logs	49
Roots using logs	50

ALGEBRA

6 Linear equations — 55

How to 'move' quantities in equations	56
Linear equations with brackets	57
Linear equations with fractions	57
Cross-multiplying in fractional equations	58
Limitations of cross-multiplying	59
Some simple fractional equations	60

7 Applications of formulae in physics — 61

Mechanics	62
Linear motion with uniform acceleration	62
Forces; weight	63
Momentum and force	64
Work and potential energy	65
Kinetic energy	67
Transfer of energy	67
Electricity	68
Current; potential difference; resistance	68
Charge and electrolysis	69
Series and parallel resistances; resistivity	70
Electrical energy	72

Heating effect of current 72
Electrical power 73

Heat 74
Heat capacity 74
Specific heat capacity 75
Electrical heating 76
Specific latent heat of vaporization (evaporation) 76
Specific latent heat of fusion 77

Geometrical optics 78
Curved mirrors and lenses 78
Curved mirrors and f 79
Object and image calculations 79
Lenses and f 81
Object and image calculations with lenses 81

8 Changing formulae: mechanics; electricity; heat 85

Mechanics 85
Linear equation 85
Equation with squares 86
Equation with square root 87
Exploding objects 88
Colliding objects 88

Electricity 90
Current; p.d.; resistance 90
Resistivity 92
E.m.f. and internal resistance 93

Heat 95
Heat capacity; specific heat capacity 95
Heat exchanges 96
Specific latent heat 97
Latent heat of fusion 99

9 Simultaneous and quadratic equations 100

Simultaneous equations 100
Elimination method 100
Substitution method 101
Magnification in lenses or curved mirrors 102

Quadratic equations 104
Solving by factors 104
Solving by formula 105

CONTENTS

10 Graphs. Linear graphs and applications — 107

Types of graphs — 107
Histogram; frequency curve — 108
Continuous graphs — 109

Linear graphs and applications in physics — 116
Linear graphs passing through origin — 117
Linear graphs not passing through origin — 117
Positive and negative gradients — 117
Some linear graphs in physics — 118
Elasticity: Hooke's Law — 118
Gases: Boyle's Law — 119
Electrical resistance — 121
Pendulum method for g — 121

11 Proportional relationships — 125

Resistance — 125
Gases — 126
Changing gas volumes to s.t.p. — 127
Square-law relationships — 127
Electrical heating — 128
Inversely-proportional relationships — 130
Boyle's Law for gases — 130
Wavelength and frequency of waves — 131
Inverse-square relationships — 132
Resistance and diameter of cross-section — 133
Law of gravitation — 134

12 Chemistry calculations: reactions; electrolysis — 136

Masses and volumes in reactions — 136
Masses — 136
Volumes of gases — 138

Electrolysis — 141
Facts about the mole — 141
Relative atomic mass; molar mass — 141
Electrolysis — 141

TRIGONOMETRY

13 Basic trigonometry — 147

Sine and applications in optics — 147
Sine of angle — 147
Values of sines — 148
Sines of 30°, 45°, and 60° — 149

Sine law of refraction 149
Calculations on refraction 150
Critical angle 151
Calculation of critical angle 152

Cosine and applications in mechanics 154
Cosine of angle 154
Cosine values for 30°, 45°, and 60° 155
Components of forces 155
Magnitude of component 156
Forces in equilibrium 157

Tangent and applications in vectors 159
Tangent of angle 159
Tangents of 60°, 30°, and 45° 160
Perpendicular vectors; resultant of velocities 161
Forces and their resultants 162

Revision papers 165

Answers 172

Preface

Numeracy and mathematical formulae are essential for a fuller understanding of many science topics at the Ordinary level stage. This book provides a straightforward concise course in basic mathematics for science students taking Ordinary level examinations or those of a similar standard in CSE or technical examinations. It deals with mathematics as a tool; it is not concerned with mathematical concepts, which are discussed in specialist books on the subject. The book does not replace any science textbooks, but equipped with the necessary basic mathematical skills, all students should be able to proceed with confidence to a sixth-form study of any science subject.

The text begins with a revision of basic arithmetic such as fractions, decimals and logarithms. It then continues with the algebra required for O-level science topics, such as linear equations, transformation of formulae, graphs, proportional relationships, and concludes with basic trigonometry. Throughout the text numerous O-level physics topics have been discussed to show how this basic mathematics is used, together with a selection of chemistry calculations. Exercises at the end of each section, and revision papers at the end of the book, are provided.

The book can be used in any order, and for any suitable age or examination group, as the teacher wishes. No doubt teachers will expand topics or exercises as they think necessary, or add other topics to the selection given. Students familiar with the basic mathematics can omit the text concerned but are advised to do the exercises. For a full background discussion of the O-level physics formulae used in this book, with which we assume the reader is familiar, reference should be made to the author's *Principles of Physics* (Hart-Davis). For the topics in chemistry, reference should be made to *New Certificate Chemistry* by Holderness and Lambert (Heinemann).

The basic mathematics in the book should also be useful for technicians in the services and in industry. Students who require a further study of mathematical skills for A-level physics or chemistry are referred to *The Mathematics of Physics* by Avery and Nelkon (Heinemann).

The author is indebted to Dr. I. L. Finar, The Polytechnic of North London, Mrs. M. Pond, The Mount School for Girls, London, and S. S. Alexander, formerly Woodhouse School, London, for their generous assistance with topics. He is also grateful to J. H. Avery, formerly of Stockport Grammar School, C. F. Tolman, Whitgift School, Croydon, and D. E. Armit, formerly of William Ellis School, London, for their constructive suggestions.

The author acknowledges with thanks the kind permission of the following Examining Boards to reproduce past questions:

University of London University Entrance and School Examinations Council (L), Oxford Delegacy of Local Examinations (O).

ARITHMETIC

1 Fractions: Adding, Subtracting, Multiplying, Dividing

Figure 1.1.

Adding and subtracting fractions

Fractions are common in everyday life. We can add or subtract fractions by changing all the fractions to the same, or 'common', denominator (the 'denominator' is the number at the bottom of the fraction, the 'numerator' is the number at the top of the fraction). For easy calculation, the common denominator should be the lowest one. For example, if the denominators were 2, 3 and 4 respectively, the lowest common denominator would be 12—it is the smallest number into which 2, 3 and 4 will divide. We illustrate this by examples.

Examples

1. Find $\frac{3}{4} + \frac{5}{12} - \frac{5}{6}$

The lowest common denominator for the denominators 4, 12 and 6 is 12. So we change each fraction to one with a denominator of 12. Then

$$\frac{3}{4} + \frac{5}{12} - \frac{5}{6} = \frac{9}{12} + \frac{5}{12} - \frac{10}{12}$$

$$= \frac{9 + 5 - 10}{12}$$

$$= \frac{4}{12} = \frac{1}{3}$$

2. Find $3\frac{2}{5} - 1\frac{7}{12} + 2\frac{3}{10}$

Firstly, adding and subtracting the whole numbers, we have $3 - 1 + 2 = 4$.

Secondly, the denominators 5, 12 and 10 have a lowest common denominator of 60. So we change each fraction to one with a denominator of 60. Then

$$3\frac{2}{5} - 1\frac{7}{12} + 2\frac{3}{10} = 3\frac{24}{60} - 1\frac{35}{60} + 2\frac{18}{60}$$

$$= 4\frac{24 - 35 + 18}{60}$$

$$= 4\frac{7}{60}$$

EXERCISE 1

Calculate

1. $\frac{7}{10} + \frac{3}{5} - \frac{3}{4}$

2. $1\frac{1}{4} - 2\frac{3}{8} + 3\frac{1}{12}$

3. $3\frac{3}{5} - 1\frac{19}{20}$

4. $2\frac{2}{5} + \frac{2}{3} - 1\frac{3}{4}$

5. $4\frac{2}{3} - 1\frac{5}{9} - \frac{5}{6}$

FRACTIONS: ADDING, SUBTRACTING, MULTIPLYING, DIVIDING

6. A silver coin is an alloy by mass of silver $\frac{1}{2}$, copper $\frac{2}{5}$, nickel $\frac{1}{20}$, and zinc as the remaining metal (*see* Figure 1.2). What fraction of zinc is used?

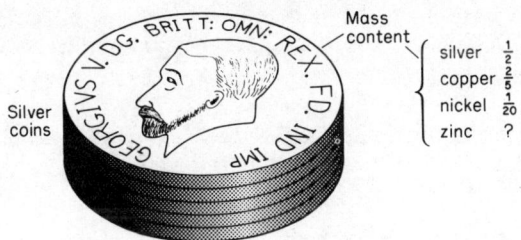

Figure 1.2.

7. Find the mass of an alloy which has $\frac{3}{20}$ kg nickel, $\frac{1}{8}$ kg iron, and $\frac{19}{40}$ kg chromium.

8. A mixture of gases by volume consists of oxygen $\frac{1}{4}$, nitrogen $\frac{2}{5}$, carbon dioxide $\frac{3}{10}$, and the rest air. What fraction of air is present?

9. Calculate the mass of a cake if its ingredients are $\frac{3}{8}$ kg flour, $\frac{1}{5}$ kg sugar, $\frac{1}{10}$ kg eggs and $\frac{17}{40}$ kg butter.

Multiplying fractions

If fractions are multiplied together, first simplify by dividing the top (numerator) and bottom (denominator) of *any* of the fractions by a common factor, if there is one. For example, in

$$\frac{4}{7} \times \frac{21}{32}$$

(i) divide 4 (top) into 32 (bottom), which gives 8, and (ii) divide 7 (bottom) into 21 (top), which gives 3. So

$$\frac{4}{7} \times \frac{21}{32} = \frac{\cancel{4}^1}{\cancel{7}_1} \times \frac{\cancel{21}^3}{\cancel{32}_8}$$

Now multiply the top numbers, which gives 3, and multiply the bottom numbers, which gives 8. So the result is the fraction $\frac{3}{8}$.

A 'mixed' number, that is, a whole number and fraction, must first be changed completely to a fraction; for example

$$2\frac{3}{4} = \frac{11}{4}, \quad 1\frac{2}{3} = \frac{5}{3}, \quad 3\frac{1}{7} = \frac{22}{7}$$

Examples

1. Find $\frac{3}{16} \times \frac{4}{15} \times \frac{10}{21}$

We note that (i) 3 (top) divides into 21 (bottom); that 5 divides into 10 (top) and into 15 (bottom); and that 4 (top) divides into 16 (bottom).

Simplifying,

$$\frac{\cancel{3}^{\,1}}{\cancel{16}_{\,4\;2}} \times \frac{\cancel{4}^{\,1}}{\cancel{15}_{\,3}} \times \frac{\cancel{10}^{\,2\;1}}{\cancel{21}_{\,7}} = \frac{1 \times 1 \times 1}{2 \times 3 \times 7} = \frac{1}{42}$$

2.
$$2\tfrac{2}{3} \times 1\tfrac{5}{16} = \frac{\cancel{8}^{\,1}}{\cancel{3}_{\,1}} \times \frac{\cancel{21}^{\,7}}{\cancel{16}_{\,2}} = \frac{7}{2} = 3\tfrac{1}{2}$$

3.
$$1\tfrac{5}{9} \times 2\tfrac{2}{5} \times 3\tfrac{1}{7} = \frac{\cancel{14}^{\,2}}{\cancel{9}_{\,3}} \times \frac{\cancel{12}^{\,4}}{5} \times \frac{22}{\cancel{7}_{\,1}} = \frac{176}{15} = 11\tfrac{11}{15}$$

EXERCISE 2

Simplify:

1. $\frac{4}{15} \times \frac{3}{14}$
2. $\frac{5}{6} \times \frac{5}{6}$
3. $3\tfrac{1}{3} \times 1\tfrac{2}{5}$
4. $2\tfrac{1}{2} \times 5\tfrac{1}{3}$
5. $3\tfrac{3}{5} \times 3\tfrac{1}{3} \times 1\tfrac{2}{7}$
6. $2\tfrac{1}{4} \times 1\tfrac{3}{11} \times 3\tfrac{1}{7}$

7. Find the distance travelled by a car in $\tfrac{5}{6}$ h if its average speed is $40\tfrac{1}{2}$ km/h (distance = speed × time).

8. A tank contains $2\tfrac{1}{3}$ gal of water (Figure 1.3). Find the volume of water in litres if 1 gal = $4\tfrac{1}{2}$ litres.

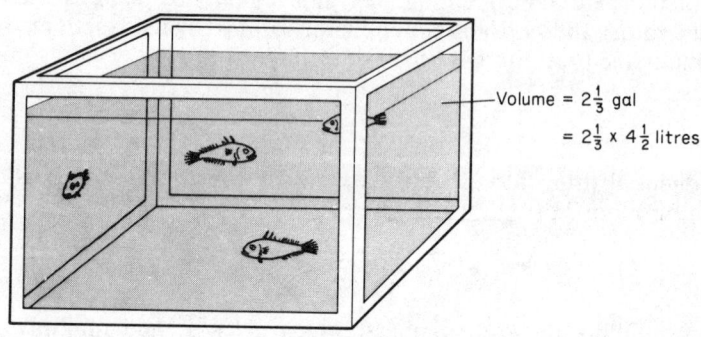

Figure 1.3.

9. The content by mass of an alloy is copper $\tfrac{2}{5}$, nickel $\tfrac{3}{8}$, and the rest manganese. Calculate the mass of each metal in $1\tfrac{1}{2}$ kg of the alloy.

10. A bicycle pedal is pushed by a force of $7\tfrac{1}{2}$ N (newton) at a perpendicular distance of $\tfrac{3}{20}$ m from its axle. What is the turning effect or *moment* of the force about the axle? (Moment = force × perpendicular distance, and unit of moment is 'N m'.)

FRACTIONS: ADDING, SUBTRACTING, MULTIPLYING, DIVIDING 7

11. The dimensions of a carpet are $10\frac{1}{2}$ ft $\times 12\frac{1}{2}$ ft. If 1 ft $= \frac{3}{10}$ m approximately, (i) calculate each dimension in metres, (ii) find the area of the carpet in metre^2 using your answers in (i).

12. Calculate in kilograms (kg) the mass of a $7\frac{1}{2}$ lb baby, given that 1 lb $= \frac{9}{20}$ kg.

Dividing fractions

If 1 is divided by $\frac{1}{5}$, the result is 5. We can see that the result is the same as multiplying 1 by $\frac{5}{1}$, that is, by the fraction turned upside down. So

$$1 \div \tfrac{1}{5} = 1 \times \tfrac{5}{1} = 5$$

Similarly,

$$1 \div \tfrac{2}{5} = 1 \times \tfrac{5}{2} = \tfrac{5}{2} = 2\tfrac{1}{2}$$

Generally, in dividing by fractions, (i) change the $\div$ sign to a $\times$ sign, and (ii) then turn upside down the fraction following the $\div$ sign. So

1. $\dfrac{5}{6} \div \dfrac{25}{36} = \dfrac{\cancel{5}^1}{\cancel{6}_1} \times \dfrac{\cancel{36}^6}{\cancel{25}_5} = \dfrac{6}{5} = 1\dfrac{1}{5}$

2. $1\dfrac{3}{4} \div 4\dfrac{2}{3} = \dfrac{7}{4} \div \dfrac{14}{3} = \dfrac{\cancel{7}^1}{4} \times \dfrac{3}{\cancel{14}_2} = \dfrac{3}{8}$

3. $\dfrac{1}{1\frac{2}{3}} = 1 \div 1\dfrac{2}{3} = 1 \div \dfrac{5}{3} = 1 \times \dfrac{3}{5} = \dfrac{3}{5}$

4. $\dfrac{2}{3\frac{1}{7}} = \dfrac{2}{\frac{22}{7}} = \dfrac{2 \times 7}{22} = \dfrac{7}{11}$

5. $1\dfrac{1}{4} \times 1\dfrac{2}{3} \div 6\dfrac{1}{4} = \dfrac{5}{4} \times \dfrac{5}{3} \div \dfrac{25}{4}$

$= \dfrac{\cancel{5}^1}{\cancel{4}_1} \times \dfrac{\cancel{5}^1}{3} \times \dfrac{\cancel{4}^1}{\cancel{25}_1} = \dfrac{1}{3}$

Example

A cyclist travels a distance of $6\frac{1}{4}$ miles in $\frac{1}{4}$ h. If 1 mile $= 1\frac{3}{5}$ km, calculate the average speed in km/h (Figure 1.4).

$$6\tfrac{1}{4} \text{ miles} = 6\tfrac{1}{4} \times 1\tfrac{3}{5} \text{ km}$$

$$\therefore \text{ average speed} = \dfrac{\text{distance}}{\text{time}} = \dfrac{6\frac{1}{4} \times 1\frac{3}{5} \text{ km}}{\frac{1}{4} \text{ h}}$$

$$= \dfrac{\cancel{25}^5}{\cancel{4}_1} \times \dfrac{8}{\cancel{5}_1} \times \dfrac{\cancel{4}^1}{1} \text{ km/h}$$

$$= 40 \text{ km/h}$$

Figure 1.4.

EXERCISE 3

Calculate

1. $\frac{2}{3} \div \frac{4}{5}$

2. $\frac{3}{4} \div 1\frac{7}{9}$

3. $\dfrac{1}{2\frac{2}{5}}$

4. $1\frac{1}{2} \div 2\frac{7}{12}$

5. $\dfrac{2}{2\frac{3}{4}}$

6. $4\frac{1}{6} \div 5$

7. $2\frac{1}{3} \times \frac{9}{14} \div 2\frac{1}{4}$

8. $\dfrac{1}{3\frac{1}{3}} \div \dfrac{9}{25}$

9. A bus travels $37\frac{1}{2}$ miles in $\frac{3}{4}$ h. If $1\,\text{km} = \frac{5}{8}$ mile approximately, calculate (i) the distance travelled in km, (ii) the average speed in m.p.h. (average speed = distance/time), (iii) the average speed in km/h.

10. How many rectangular stones, each $1\frac{1}{2}\,\text{m} \times 1\frac{3}{4}\,\text{m}$, can be laid in an area of $84\,\text{m}^2$?

11. (i) A 40 litre tank is completely filled with liquid. What volume in gallons is used, assuming $1\,\text{gal} = 4\frac{1}{2}$ litres approximately? (ii) Nine-tenths of a 45 litre tank contains liquid. What is the liquid volume in gallons?

12. The dimensions of a container is $1\frac{1}{3}\,\text{m} \times 1\frac{1}{2}\,\text{m} \times 1\frac{1}{8}\,\text{m}$. How many containers can be loaded on a truck whose volume capacity is $36\,\text{m}^3$?

FRACTIONS: ADDING, SUBTRACTING, MULTIPLYING, DIVIDING 9

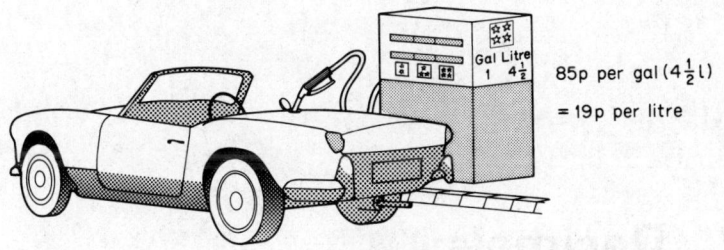

Figure 1.5.

2 Decimals: Adding, Subtracting, Multiplying, Dividing

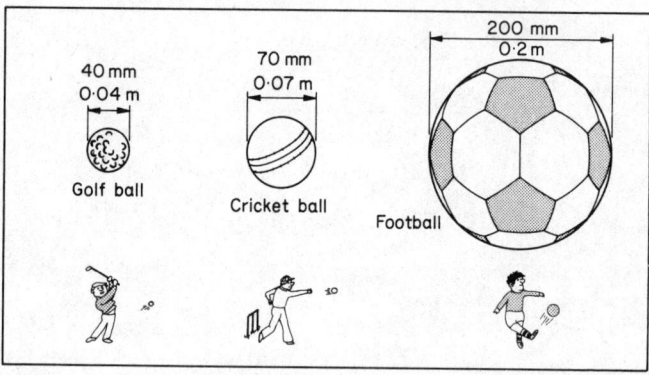

Figure 2.1.

Decimals and fractions

The metric system, used in all branches of science, is based on multiples of 10 such as 10, 100, 1000 and so on, and submultiples of 10 such as $\frac{1}{10}$, $\frac{1}{100}$, $\frac{1}{1000}$ and so on.

For convenience, fractions are written as *decimals*; these are numbers with a point in front of the number. Thus

$$\frac{1}{10} = 0.1, \qquad \frac{1}{100} = 0.01, \qquad \frac{1}{1000} = 0.001$$

We write the number or digit '0' in front of the decimal points to show there are no whole numbers, since fractions are less than 1. In contrast, we write

$$1\frac{7}{10} = 1.7, \qquad 2\frac{19}{100} = 2.19, \qquad 3\frac{7}{1000} = 3.007$$

You should note that the number of digits (figures) following the decimal point is equal to the number of noughts in the denominators

DECIMALS: ADDING, SUBTRACTING, MULTIPLYING, DIVIDING

of the fractions; $\frac{1}{10}$ has one nought, $\frac{1}{100}$ has two noughts, and so on. So changing from *decimals to fractions*, we write

$$0.3 = \tfrac{3}{10}, \qquad 1.06 = 1\tfrac{6}{100} = 1\tfrac{3}{50}, \qquad 2.005 = 2\tfrac{5}{1000} = 2\tfrac{1}{200}$$

Examples

Change 0.45, 1.025 and 0.0004 to their simplest fractions.

1. $\qquad 0.45 = \tfrac{45}{100} = \tfrac{9}{20}$
2. $\qquad 1.025 = 1\tfrac{25}{1000} = 1\tfrac{5}{200} = 1\tfrac{1}{40}$
3. $\qquad 0.0004 = \dfrac{4}{10\,000} = \dfrac{1}{2500}$

EXERCISE 4

Change the following decimals to their simplest fractional values (in Questions 4, 7, 8, 10 leave the whole numbers unaltered).

1. 0.6 2. 0.04 3. 0.006 4. 9.5 5. 0.005
6. 0.004 7. 1.08 8. 2.2 9. 0.0008 10. 5.23

Metric system and SI units

The metric system uses powers of 10 in units of length and mass, for example.

Length In the SI system (SI = Système International), now adopted for international use in science, the metre (m) is the unit of length. Other units are:

$$1 \text{ decimetre (dm)} = \tfrac{1}{10}\text{ m} = 0.1\text{ m}$$
$$1 \text{ centimetre (cm)} = \tfrac{1}{100}\text{ m} = 0.01\text{ m}$$
$$1 \text{ millimetre (mm)} = \tfrac{1}{1000}\text{ m} = 0.001\text{ m}$$
$$1 \text{ kilometre (km)} = 1000\text{ m}$$

Approximate equivalents are: 1 inch = 2.5 cm = 25 mm; 1 ft = 30 cm = 300 mm; 1 yd = 0.9 m; 1 mile = 1.6 km.

The diameter of a golf ball is about 40 mm (4 cm or 0.04 m); the diameter of a cricket ball is about 70 mm (7 cm or 0.07 m); the diameter of a soccer football is about 200 mm (20 cm or 0.2 m) as shown in Figure 2.1. A speed of 30 m.p.h. = 48 km/h (approximately).

Area The m^2 is the SI unit of area. Other units are: mm^2, cm^2, dm^2, hectare.

$$1 \text{ cm}^2 = (10 \text{ mm})^2 = 100 \text{ mm}^2$$
$$1 \text{ m}^2 = (1000 \text{ mm})^2 = 1\,000\,000 \text{ mm}^2$$
$$1 \text{ hectare} = 10\,000 \text{ m}^2$$

Approximate equivalents are: 1 square yard (yd^2) = 0.8 m^2; 1 acre = 0.4 hectare. A football pitch measuring 110 yd × 66 yd has an area of about $1\frac{1}{2}$ acre or 0.6 hectare.

Volume The m^3 is the SI unit of volume. Other units are cm^3, litre (l), dm^3,

$$1 \text{ litre} = 1000 \text{ cm}^3$$
$$1 \text{ dm}^3 = (10 \text{ cm})^3 = 1000 \text{ cm}^3 = 1 \text{ l}$$

1 pint is slightly greater than $\frac{1}{2}$ l; 1 gallon is about $4\frac{1}{2}$ l.
At s.t.p. (0°C and 760 mm Hg pressure), the volume of any gas is about 22.4 dm^3 or 22.4 l. From above,

$$22.4 \text{ dm}^3 = 22\,400 \text{ cm}^3$$

Mass The SI unit of mass is the kilogram (kg).

$$1 \text{ gram (g)} = \frac{1}{1000} \text{ kg} = 0.001 \text{ kg}$$

Approximate equivalents are: 1 oz = 28 g; 4 oz = 112 g; 1 lb = 454 g = 0.454 kg; 1 ton = 1000 kg.

Adding and subtracting decimals

The addition or subtraction of decimals should be done with the decimal points *in line* with each other. In this way the whole numbers and decimal (fraction) parts are in line.

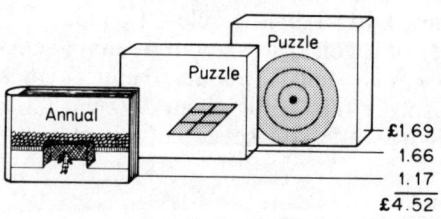

Figure 2.2.

DECIMALS: ADDING, SUBTRACTING, MULTIPLYING, DIVIDING

Examples

1. Add 4.07 m, 5.4 m, 2.75 m

$$\begin{array}{r} 4.07 \text{ m} \\ +\ 5.4 \\ +\ 2.75 \\ \hline 12.22 \text{ m} \end{array}$$

2. Subtract £4.58 from £10.13

$$\begin{array}{r} £10.13 \\ -\ 4.58 \\ \hline £5.55 \end{array}$$

(In subtraction, always *check* by adding the bottom two lines, which should total the top line, 10.13.)

3. Subtract 4.96 kg from 12.2 kg

Here we can write 12.2 as 12.20

$$\begin{array}{r} 12.20 \text{ kg} \\ -\ 4.96 \\ \hline 7.24 \text{ kg} \end{array}$$

Significant figures

A measurement of length such as 4.26 m shows a degree of accuracy of measurement of 0.01 m in 4.26 m, which is 1 in 426. Expressed in centimetres as 426 cm, the length has the same degree of accuracy, which is 1 in 426. Expressed as 4260 mm to the nearest 10 mm, this is an accuracy of 10 in 4260 or 1 in 426, the same as before.

The figures in a measurement which show the degree of accuracy are called the *significant figures*. So

496.2	is significant to 4 figures (or 4. s.f.)
3.60	is significant to 3 figures (or 3 s.f.)—the nought at the end is significant
1.8	is significant to 2 figures
5000	is significant to 1 figure (the noughts after the first figure are not significant, that is, '5260 to 1 s.f.' is 5000)

A number such as 2.44 is 2.4 to 2 s.f., but 2.45 is given as 2.5 to 2 s.f. Similarly, 3.75 is 3.8 to 2 s.f., but 3.74 is 3.7 to 2 s.f.

In *decimals* or numbers less than 1, the noughts after the decimal point are not counted as significant figures. So 0.0062 is significant to 2 figures; 0.03 is significant to 1 figure, and so is 0.00009. If we add 3.42 to 2.8, the result is 6.22 to 3 s.f., 6.2 to 2 s.f. and 6 to 1 s.f.

Measurements and significant figures

The length of a rod A measured with a ruler graduated in millimetres may be stated as 213 mm or 21.3 cm. The *accuracy* of the measurement is 1 in 213 or about 1 in 200 in round figures. The measurement of 21.3 cm implies that the actual length of A may lie between 21.25 and 21.35 cm, but we cannot guarantee the accuracy of the second decimal place using our ruler. So the measurement is given as 21.3 cm to 3 s.f.

If two lengths with a different number of significant figures are added or subtracted, for example, 3.4 m and 1.26 m, the result is usually expressed to the same number of significant figures as the *least* accurate. This is 3.4 m, which has two significant figures. So

$$3.4 + 1.26 = 4.66 \text{ m} = 4.7 \text{ m} \quad (2 \text{ s.f.})$$

and

$$3.4 - 1.26 = 2.14 \text{ m} = 2.1 \text{ m} \quad (2 \text{ s.f.})$$

EXERCISE 5

Find the values of the following, giving the final answer to the number of significant figures (s.f.) in the brackets beside each question.

1. $3.6 + 7.4$ (3 s.f.)
2. $1.3 + 8.67 + 3.452$ (3 s.f.)
3. $5.12 - 1.97$ (2 s.f.)
4. $16.05 - 12.384$ (2 s.f.)
5. $4.59 + 6.74 + 3.98$ (3 s.f.)
6. $15.2 - 12.75$ (2 s.f.)
7. $3.6 + 9.52 + 7.66$ (2 s.f.)
8. $10.025 - 7.82 - 1.18$ (3 s.f.)
9. $18.2 + 3.64 - 2.85 - 7.99$ (3 s.f.)
10. $1.006 + 3.075 - 2.842$ (3 s.f.)

11. A number of articles were bought for £0.69, £1.85, £2.64, £3.06, £4.79 and a deduction was made of £1.47 and £1.68 for returned articles. What was the net amount spent?

12. Masses of 3.64, 4.5, 7.76, and 12.89 kg of uranium are delivered to a nuclear reactor and masses of 0.75, 1.24, and 1.39 kg of used uranium are removed. What is the net mass of uranium delivered to the reactor?

13. Four lengths of wood are respectively 1.89 m, 2.64 m, 3.5 m, and 6.58 m. If lengths of 1.68 m and 2.76 m were cut off, what is the total length of wood remaining?

Multiplying decimals

1. Suppose we wish to find the result of 0.2×0.6. Now

$$0.2 = \tfrac{2}{10} \quad \text{and} \quad 0.6 = \tfrac{6}{10}$$

DECIMALS: ADDING, SUBTRACTING, MULTIPLYING, DIVIDING

So
$$0.2 \times 0.6 = \tfrac{2}{10} \times \tfrac{6}{10} = \tfrac{12}{100} = 0.12$$

From the fractions $\tfrac{2}{10}$ and $\tfrac{6}{10}$ we see that to find 0.2×0.6, (i) multiply the numbers 2 and 6 *without* using the decimal points—this is 12; then (ii) fix the decimal point *two places from the right*—the number of places is the sum of the figures after the decimal point in the two decimals multiplied, that is, one for 0.2 and one for 0.6. So the answer is 0.12.

2. Now consider 1.3×0.02. Follow the same rule. So (i) multiply 13×2—the answer is 26, (ii) the sum of the figures beyond the decimal points is $1+2=3$. So starting from 6 and counting 3 places to the left, the final answer is 0.026. The nought is needed in front of the 2 to make the 3 figures required after the decimal point.

3. To calculate 4.64×0.0072, (i) multiply 464×72—the answer is 33 408; (ii) the sum of the figures after the decimal points is $2+4=6$. So count six places from the end figure 8 to fix the decimal—the answer is 0.033 408 (5 figures in 33408 requires an extra 0 to make 6 figures).

Check Where possible, make a rough check on the calculation. For example, in 41.3×2.9, the answer is roughly $41 \times 3 = 123$. So an answer such as 118.7 is possible but 1187.0 is impossible. The rough check helps to settle the position of the decimal point in some calculations.

Order of accuracy

In multiplication or division of two measurements, we must be careful to give the result to a correct order of accuracy.

Suppose the measured lengths of the sides of a rectangle are stated as 1.1 m and 1.2 m respectively. Each length is then given to an accuracy of about 1 in 10. Now, by *calculation*,

$$\text{area of rectangle} = 1.1 \times 1.2 = 1.32 \text{ m}^2$$

The figure 1.32 implies an accuracy of about 1 in 100 in round figures. But the individual lengths had an accuracy of about 1 in 10. Since the result for the area cannot be more accurate than the least accurate measurement, we give the result as 1.3 m^2. This is about the same order of accuracy as the two lengths.

In an experiment, the final result must not be more accurate than the least accurate measurement. In Figure 2.3, for example, the resistance R of a wire is measured by a voltmeter–ammeter method (p. 69). The voltmeter reads a value V of 2.6 V to two significant figures and the ammeter a current value I of 1.1 A. Now $R = V/I$. So,

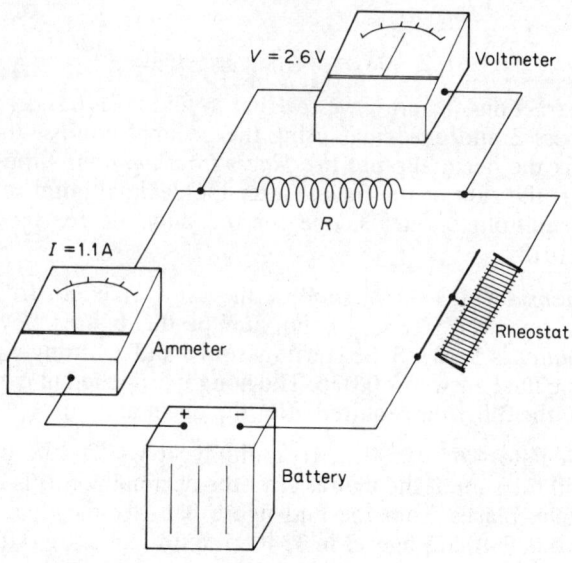

Figure 2.3.

by calculation,

$$R = \frac{V}{I} = \frac{2.6}{1.1} = 2.364 \text{ ohms}$$

The result for R is given as 2.4 ohms, about the same accuracy as the individual measurements. An answer of 2.364 ohms would be wrong, as this would imply an accuracy of 1 in 2364 or about 1 in 2000, well beyond the accuracy of the two individual measurements.

In general, when two measurements have a different number of significant figures, for example, 1.4 m and 2.62 m, the result obtained by multiplying or dividing them is usually given to the same number of significant figures as the *least* accurate measured value. So in this case,

$$1.4 \text{ m} \times 2.62 \text{ m} = 3.668 \text{ m}^2 \text{ by calculation} = 3.7 \text{ m}^2,$$

since 1.4 m (2 significant figures) is the less accurate value. Also

$$\frac{2.62}{1.4} = 1.871 \text{ by calculation} = 1.9,$$

since 1.9 is the result to 2 significant figures.

DECIMALS: ADDING, SUBTRACTING, MULTIPLYING, DIVIDING

EXERCISE 6

Calculate the following to the number of significant figures (s.f.) indicated in brackets:

1. 1.2×0.1 (1 s.f.)
2. 0.008×3.6 (2 s.f.)
3. 6.7×8.5 (2 s.f.)
4. 41.5×12.2 (3 s.f.)
5. 2.07×0.85 (2 s.f.)
6. 0.09×12.78 (2 s.f.)
7. 0.005×0.64 (1 s.f.)
8. 0.02×0.006 (2 s.f.)
9. 1.02×0.6 (2 s.f.)
10. 0.0004×3.05 (2 s.f.)
11. 42.6×2.1 (3 s.f.)
12. 3.84×12.6 (3 s.f.)
13. A cricket ball has a mass of 5.5 oz. If 1 oz = 0.03 kg, calculate the mass in kg (2 s.f.).
14. The volume capacity of a truck is $6.6 \, \text{m} \times 2.5 \, \text{m} \times 2.8 \, \text{m}$. Calculate the volume (3 s.f.).

Figure 2.4.

15. A man paints a house using 2.8 gal of paint. If 1 gal = 4.5 litres, calculate the volume of paint used in litres. Calculate the area covered if 1 litre covers $12 \, \text{m}^2$.
16. The volume of a sample of steel is $9.5 \, \text{cm}^3$ and the density of steel is $7.8 \, \text{g/cm}^3$. Find the mass of the sample (mass = volume × density).
17. Find the distance travelled by a car in 0.75 h at an average speed of 46.5 km/h.

18. A length of dress material is 3.8 yd and its width is 1.5 yd. Find the area of the material in square metre (m^2) if 1 yd = 0.9 m (2 s.f.).

19. Assuming 1 m.p.h. = 1.6 km/h, what is the speed of a car in km/h when it travels at (i) 28 m.p.h., (ii) 47 m.p.h. (2 s.f.).

20. A telephone box has dimensions 2.9 ft × 2.9 ft × 7.7 ft. If 1 ft = 0.3 m, calculate (i) each dimension in metres, (ii) the volume of the telephone box in m^3 (2 s.f.).

Dividing decimals

1. Suppose we need to find $0.48 \div 0.014$, that is, $\frac{0.48}{0.014}$. To clear *both* decimals and obtain whole numbers, multiply the top and bottom of the fraction by 1000. Then

$$\frac{0.48}{0.014} = \frac{0.48 \times 1000}{0.014 \times 1000} = \frac{480}{14} = 34.3 \quad (1 \text{ dec. place})$$

The same clearance of decimals is obtained by moving the decimal points in 0.48 and 0.014 *three places to the right*—this is the same as multiplying each by 1000. The result is 480 for 0.48, adding a '0' to '48', and 14 for 0.014. Similarly, to calculate

$$\frac{0.56}{0.08}$$

move the decimal point two places to the right for each of the decimals. We then obtain

$$\frac{56}{8} = 7$$

To calculate

$$\frac{0.0032}{5.3}$$

we move the decimal point four places to the right for each decimal, adding three zeros for 5.3 because it has only one decimal place. We then have

$$\frac{32}{53\,000} = 0.0006 \quad (1 \text{ dec. place})$$

2. Suppose we need to calculate

$$\frac{0.0072 \times 4.2}{0.64}$$

The greatest number of decimal places is four, in 0.0072. So to clear the decimals in 0.0072 and 0.64, move the decimal point four

DECIMALS: ADDING, SUBTRACTING, MULTIPLYING, DIVIDING 19

places to the right in each, adding two noughts after 64. So, as a first step, we obtain

$$\frac{72 \times 4.2}{6400}$$

Finally, to clear the decimal point in 4.2, move the decimal point one place to the right in the top and bottom of the fraction—this means adding a '0' to 6400. So we obtain

$$\frac{72 \times 42}{64\,000} = \frac{3024}{64\,000} = 0.0047 \quad (2 \text{ s.f.})$$

Note that we move the decimal point to the right for one pair of numbers at a time in the top and bottom. So

$$\frac{0.036 \times 0.64}{2.8} = \frac{36 \times 0.64}{2800} = \frac{36 \times 64}{280\,000} = 0.0082 \quad (2 \text{ s.f.})$$

EXERCISE 7

Calculate the following:

1. $\dfrac{0.48}{1.2}$ 2. $\dfrac{0.063}{0.9}$ 3. $\dfrac{0.32}{0.016}$

4. Find the number of parcels, each of volume 0.05 m^3, which can fill a box of volume 1.5 m^3.

5. $\dfrac{0.25 \times 0.44}{2.2}$ 6. $\dfrac{1.5 \times 3.2}{4.8}$

7. Rectangular tiles, each 0.3 m × 0.25 m, are laid on a floor of area 16.5 m^2. How many tiles are used?

8. The length of a rectangular bar of area 15.36 cm^2 is 64 cm. Find the width of the bar.

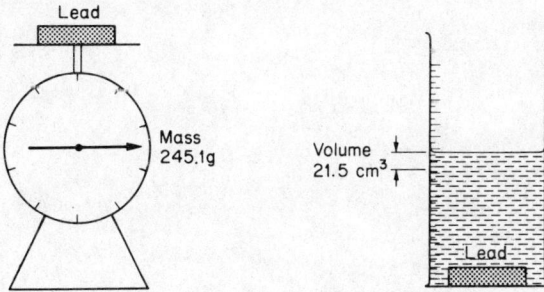

Figure 2.5.

BASIC MATHEMATICS FOR SCIENCE

9. The mass of a lump of lead is 245.1 g and its volume is 21.5 cm^3. Calculate the density of lead. (Density = mass/volume, and unit is 'g/cm^3' in this case.)

10. The heat capacity of a metal can is 442 J/K. If the mass of the can is 0.85 kg, calculate the heat capacity per kg of the metal. (Heat capacity per kg = heat capacity/mass in kg, and unit is 'J/kg K'.)

11. Calculate $\dfrac{2.4 \times 0.6}{0.036}$

12. Calculate $\dfrac{0.02 \times 0.08}{0.4}$

13. Calculate $\dfrac{3.5 \times 0.014}{0.07 \times 0.02}$

14. Calculate $\dfrac{0.025 \times 0.8}{1.6 \times 0.05}$

3 Ratios and percentages. Chemistry calculations

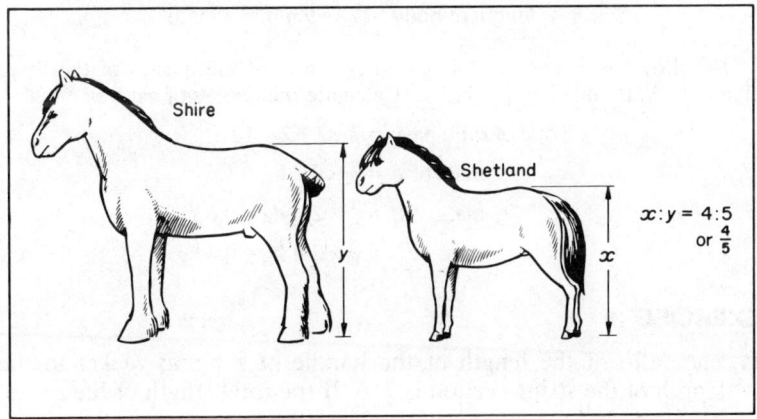

Figure 3.1.

Ratios

The *ratio* of the average height of a Shetland pony to that of a Shire horse may be expressed either as 4:5, which means 4÷5, or as $\frac{4}{5}$ (Figure 3.1). So a ratio represents a fraction. Ratios can only be used to compare similar quantities, for example, two lengths or two areas or two masses.

Ratio calculations

Suppose the ratio of the mass of water in a fresh water melon to the mass of the remaining melon is 9:1. This means that in the *total mass* of melon, say 1.0 kg, 9 parts are water and 1 part is melon. Since there are 10 parts altogether,

$$\text{mass of water} = \tfrac{9}{10} \times 1.0 \text{ kg} = 0.9 \text{ kg},$$

and

$$\text{mass of melon} = \tfrac{1}{10} \times 1.0 \text{ kg} = 0.1 \text{ kg}$$

To find the fraction of the total mass, then, we add the ratio numbers of all the various parts, in this case $9+1$, which totals 10. This gives the *denominator* of the fraction of the total. We illustrate the method in the examples which follow.

Examples

1. The ratio of the length of the tail of a mouse to that of the rest of its body is $6:5$ and its total length is 99 mm. Find the length of the tail and the body.

Total of ratio parts $= 6+5 = 11$

$\therefore$ length of tail $= \frac{6}{11} \times 99$ mm $= 54$ mm

and

length of body $= \frac{5}{11} \times 99$ mm $= 45$ mm

2. An alloy has a mass of 2.4 kg and the ratio of the masses of the three elements A, B and C in it is $7:3:2$. Calculate the masses of each element.

Total of ratio parts $= 7+3+2 = 12$

$\therefore$ mass of A $= \frac{7}{12} \times 2.4$ kg $= 1.4$ kg

mass of B $= \frac{3}{12} \times 2.4$ kg $= 0.6$ kg

mass of C $= \frac{2}{12} \times 2.4$ kg $= 0.4$ kg

EXERCISE 8

1. The ratio of the length of the handle of a tennis racket to the length of the string section is $7:6$. If the total length of the racket is 65 cm, find the lengths of the handle and string section.

2. In one form of brass, the masses of copper and zinc used are in the ratio $7:3$. Calculate the mass of each metal in 4.0 kg of brass.

3. Invar is a metal alloy made of steel and nickel in the ratio $16:9$ by mass. (i) Calculate the mass of each metal in 2.0 kg of Invar. (ii) If 320 g of steel is used in another mass of Invar, what mass of nickel is used?

4. An alloy has a total mass of 1.8 kg and is made of three elements A, B, C in the ratio $5:3:1$ by mass. Find the mass of each element used.

5. A sum of £1.20 is shared between three children in the ratio $3:2:1$. How much did each get?

6. A mixture is made of flour, egg, and milk in the ratio $3:2:10$ by mass. Calculate the mass of each in 420 g of the mixture.

7. An alloy called Manganin, used for electrical resistance wire, is made of copper, manganese, and nickel in the ratio $21:12:1$ by mass. What mass of each metal is present in 680 g of Manganin?

RATIOS AND PERCENTAGES. CHEMISTRY CALCULATIONS 23

Percentages

A *percentage* (%) is a special form of fraction; it has a denominator of 100. So

$$7\% = \tfrac{7}{100}; \qquad 20\% = \tfrac{20}{100} = \tfrac{1}{5}; \qquad 75\% = \tfrac{75}{100} = \tfrac{3}{4}$$

Thus each of the above fractions will give the percentage figure when multiplied by 100. This is generally the case. So

$$\tfrac{1}{10} = \tfrac{1}{10} \times 100\% = 10\%$$
$$\tfrac{3}{8} = \tfrac{3}{8} \times 100\% = 37.5\%$$
$$\tfrac{5}{6} = \tfrac{5}{6} \times 100\% = 83.3\%$$

Efficiency of machines

The *efficiency* of a machine is defined as a ratio or fraction, which can be changed to a percentage by multiplying the fraction by 100:

$$\text{efficiency of machine} = \frac{\text{energy (or power) obtained}}{\text{energy (or power) supplied}} \times 100\%$$

Energy is measured in *joules*, J; power is measured in *watts*, W.

A perfect machine is one from which the energy (or power) obtained is equal to the energy (or power) supplied to the machine. In this case the efficiency would be 100%. In practice, owing to friction for example, no machine is perfect. The energy (or power) obtained from a system of pulleys, or an electric motor, is less than the amount supplied.

Suppose the efficiency of a pulley system is 70%. Then the energy obtained from it is 70% or $\tfrac{7}{10}$ of the energy supplied. So if the energy supplied is 1000 J,

$$\text{energy obtained} = \tfrac{7}{10} \times 1000 = 700 \text{ J}$$

Example

An electric motor has an efficiency of 75%. The power obtained from it when driving a lathe is 600 W. Calculate the power supplied to the motor.

$$\text{Efficiency} = 75\% = \frac{3}{4}$$

$$\therefore \quad \frac{600 \text{ W}}{\text{power supplied}} = \frac{3}{4}$$

$$\therefore \quad \text{power supplied} = \frac{4}{3} \times 600 = 800 \text{ W}$$

Check $\quad \text{Efficiency} = \dfrac{\text{power obtained}}{\text{power supplied}} = \dfrac{600 \text{ W}}{800 \text{ W}} = \dfrac{3}{4} = 75\%$

EXERCISE 9

1. Change (i) to simplest fractions 85%, 70%, 55%, 37.5%, 5%; (ii) to percentages $\frac{2}{5}$, 0.78, $\frac{5}{6}$, 0.525.

2. On the Earth's surface, the area occupied by land is about 150 million km². If this is roughly 30% of the Earth's total surface area, estimate the surface area occupied by the oceans.

3. The energy obtained by pulling a barrel up an inclined plane is 400 J and the energy used or supplied was 500 J. What is the efficiency of the arrangement?

4. The efficiency of an electric drill is 70%. Calculate the power obtained from it when the power supplied is 150 W. How much power is wasted?

5. The energy obtained from a pulley system when raising a load is 3000 J. If the efficiency of the system is 60%, what energy is (i) supplied to the system, (ii) wasted?

6. In a hydraulic press, 40 000 J of energy was obtained from the machine when 64 000 J of energy was supplied. (i) What energy was wasted? (ii) What is the efficiency of the press?

CHEMICAL CALCULATIONS

Percentage mass composition

We now calculate the percentage composition by mass of some chemical compounds.

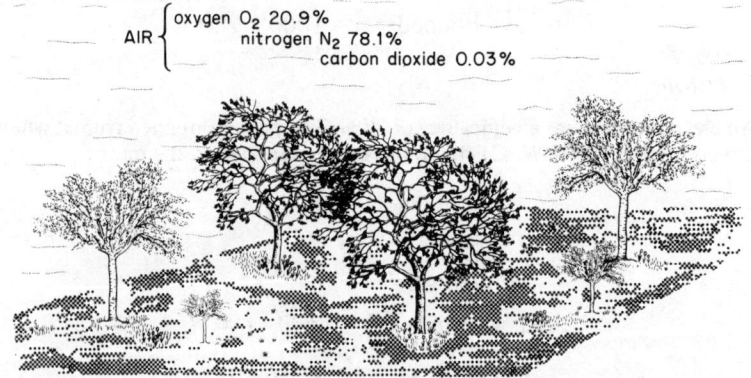

Figure 3.2.

RATIOS AND PERCENTAGES. CHEMISTRY CALCULATIONS

Examples

1. The compound CO_2 has 1 atom of carbon C and 2 atoms of oxygen O. The relative atomic masses of C and O = 12 and 16 respectively. So

   relative molecular mass of $CO_2 = 12 + (2 \times 16) = 44$

   $\therefore$ % mass of carbon C $= \frac{12}{44} \times 100\% = 27.3\%$

   % mass of oxygen O $= \frac{32}{44} \times 100\% = 72.7\%$

   Check Total % mass = 100%.

2. The compound H_2SO_4 has 2 atoms H, 1 atom S, and 4 atoms O. Given the relative atomic masses as H = 1, S = 32, and O = 16, find the percentage by mass of each element in 3.00 g of the compound.

   Relative molecular mass of $H_2SO_4 = 2 + 32 + 64 = 98$

   $\therefore$ mass of H $= \frac{2}{98} \times 3.00$ g $= 0.06$ g

   mass of S $= \frac{32}{98} \times 3.00$ g $= 0.98$ g

   mass of O $= \frac{64}{98} \times 3.00$ g $= 1.96$ g

   Check Total mass = 3.00 g.

Law of multiple proportions

This chemical law states that if two elements A and B combine to form more than one compound, then the masses of A which combine with a fixed mass of B are in a simple ratio.

Examples

1. An element X forms two oxides containing 50% and 40% of X. Show these oxides illustrate the law of multiple proportions.

   To simplify matters, suppose we have 100 g of each oxide.

   Then oxide I has 50% or 50 g X, and 50% or 50 g O

   and oxide II has 40% or 40 g X, and 60% or 60 g O.

   From oxide II, we see that 50 g of X would combine with a mass of O of

   $\frac{50}{40} \times 60$ g $= 75$ g

   But in oxide I, 50 g of X combines with 50 g of O. So the ratio of the masses of O which combine with the same mass 50 g of X is

   50 g : 75 g $= 2 \times 25$ g : 3×25 g $= 2 : 3$

2. An element X forms two oxides containing 77.5 and 69.6% respectively by mass of X. If the first oxide has the formula XO, what is the formula of the second oxide?

   Consider, for simplification, 100 g of each oxide. Then,

   for first oxide,   there is 77.5 g of X, and so 22.5 g of O

   for second oxide, there is 69.6 g of X, and so 30.4 g of O

   (1)

So in the second oxide,

77.5 g X would combine with $\dfrac{77.5}{69.6} \times 30.4$ g O = 33.8 g by calculation

From (1), we see that the ratio of the two masses of oxygen which combine with the same mass of X is $33.8 : 22.5 = 33 : 22$ in round figures $= 3 : 2$.
So if the first oxide is XO, the second oxide is $XO_{3/2}$ or X_2O_3.

EXERCISE 10

1. In the following compounds, calculate the mass percentage of the element or compound in brackets, given the relative atomic masses given at the end (answer to one decimal place):
   (i) CO (C). $C = 12$, $O = 16$
   (ii) AgCl (Cl). $Ag = 108$, $Cl = 35.5$
   (iii) Al_2O_3 (Al). $Al = 27$, $O = 16$
   (iv) H_2SO_4 (S). $H = 1$, $S = 32$, $O = 16$
   (v) $CuSO_4.5H_2O$ ($5H_2O$). $Cu = 63.5$, $S = 32$, $O = 16$

Copper sulphate crystal
$CuSO_4.5H_2O^{(blue)}$

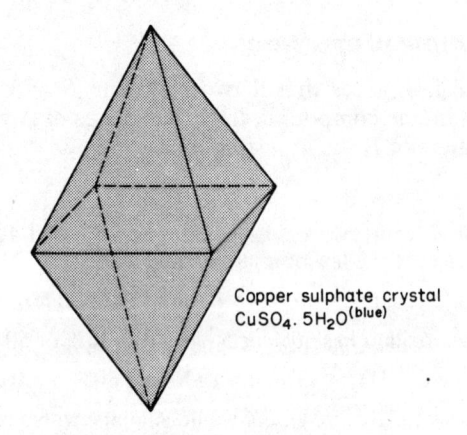

Figure 3.3.

2. A metal Y forms two oxides A and B. 1 g of A gives 0.24 g of oxygen and 1 g of B gives 0.39 g of oxygen. Show that these figures illustrate the law of multiple proportions.

3. The two chlorides of the metal X have respectively 54.1% and 37.5% of Cl. Show these figures are in agreement with the law of multiple proportions.

4. A metal X forms two oxides A and B. 1 g of A on reduction gives 0.88 g of X and 1 g of B gives 0.79 g of X. Show that these results are in agreement with the law of multiple proportions.

RATIOS AND PERCENTAGES. CHEMISTRY CALCULATIONS 27

Chemical formulae

We can find the simplest or 'empirical' formula of a chemical compound from measurements of the percentage masses of the elements in the compound and a knowledge of their relative atomic masses. This is illustrated in the following examples.

Examples

1. Calculate the empirical formula of a compound which has by mass 53.8% iron (Fe) and 46.2% sulphur (S). Relative atomic mass of Fe = 56, of S = 32.

In 100 g of compound, we have

Fe	S
53.8 g	46.2 g

Divide by relative atomic mass:

$\dfrac{53.8}{56}$	$\dfrac{46.2}{32}$
= 0.96 mol	= 1.44 mol

Divide by smaller number of moles:

$\dfrac{0.96}{0.96}$	$\dfrac{1.44}{0.96}$
= 1	= 1.5

This is the *ratio* of the number of atoms in the molecule.
But atoms must be present in whole numbers.
So the smallest number of atoms present is

1×2	1.5×2
= 2	= 3

So empirical formula of compound is Fe_2S_3

2. Find the empirical formula of a compound with the following composition:
By mass, 22.6% Zn, 11.1% S, 22.3% O, 44% water of crystallization. (Relative atomic mass of Zn = 65, of S = 32, of O = 16, relative molecular mass of water = 18.)

Suppose the compound has a mass of 100 g.

The number of mole of $Zn = \dfrac{22.6}{65} = 0.35$

of $S = \dfrac{11.1}{32} = 0.35$

of $O = \dfrac{22.3}{16} = 1.39$

of water $= \dfrac{44}{18} = 2.44$

So ratio of number of atoms and water molecules
$$= 0.35:0.35:1.39:2.44$$
$$= 1:1:4:7 \quad \text{in round figures}$$
Thus the empirical formula is $ZnSO_4.7H_2O$.

EXERCISE 11

1. Find the empirical formula of a compound which has the following composition by mass: 34.7% Fe, 65.3% Cl.
   (Relative atomic mass: Fe = 56, Cl = 35.5.)

2. A hydrocarbon has the following composition by mass: 20% H, 80% C. Find its empirical formula. If the relative molecular mass is 30, find the actual formula.
   (Relative atomic mass: H = 1, C = 12.)

3. A $CuSO_4$ hydrate has 64% $CuSO_4$ by mass. What is its empirical formula?
   (Relative atomic mass: Cu = 63.5, S = 32, O = 16, H = 1.)

4. Find the empirical formula for a compound which has the following composition by mass: 28.1% Fe, 35.7% Cl, 36.2% water of crystallization.
   (Relative atomic mass: Fe = 56, Cl = 35.5, O = 16, H = 1.)

5. The composition by mass of an organic compound is: 12.8% C, 2.1% H, 85.1% Br. If the compound has a relative molecular mass of 188, find its chemical formula.
   (Relative atomic mass: C = 12, H = 1, Br = 80.)

4 Indices

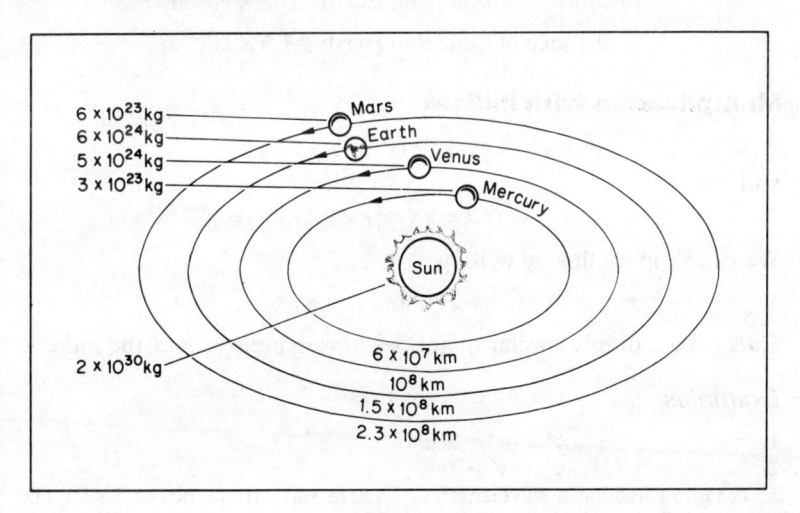

Figure 4.1.

The repeated multiplication $x \times x \times x \times x$ is written briefly as x^4; the power or *index* 4 is the number of multiplied xs. So

$$a \times a = a^2 \ (not\ 2a) \quad \text{and} \quad r \times r \times r = r^3 \ (not\ 3r)$$

So if $x = 5$, then $x^2 = x \times x = 5 \times 5 = 25$, and $x^3 = 5 \times 5 \times 5 = 125$. We see that

$$10^2 = 10 \times 10 = 100 \quad \text{and} \quad 10^4 = 10 \times 10 \times 10 \times 10 = 10\,000$$

Note that x can be written with an index as x^1.

Powers of 10

Powers of 10 are often used in science to write large numbers down concisely. For example,

$$100\,000 \text{ (hundred thousand)} = 10^5$$

and

$$1\,000\,000 \text{ (million)} = 10^6$$

So

$$3 \times 10^8 = 3 \times 100\,000\,000$$
$$= 300\,000\,000 \text{ (3 hundred million)}$$

Here are some large values expressed in powers of 10 (Figure 4.1):

$$\text{mass of Earth} = 6 \times 10^{24} \text{ kg}$$
$$\text{mass of Sun} = 2 \times 10^{30} \text{ kg}$$
$$\text{distance of Moon from Earth} = 3.8 \times 10^8 \text{ m}$$
$$\text{distance of Sun from Earth} = 1.5 \times 10^{11} \text{ m}$$

Multiplication with indices

$$x^2 \times x^3 = (x \times x) \times (x \times x \times x) = x^5$$

and

$$r^2 \times r^5 = (r \times r) \times (r \times r \times r \times r \times r) = r^7$$

We can simplify this by writing

$$x^2 \times x^3 = x^{2+3} = x^5 \quad \text{and} \quad r^2 \times r^5 = r^{2+5} = r^7$$

Rule To multiply similar quantities having indices, *add* the indices.

Examples
1. $\quad 4a^3 \times 3a^4 = 12a^7$
2. $\quad 2y^5 \times 4y^4 = 8y^9$
3. A radio wave has a wavelength of 1500 m and a frequency of 2×10^5 Hz. Calculate the speed of the wave if speed = wavelength × frequency, and the speed is in m/s when the wavelength is in m and the frequency in Hz (hertz).

$$\text{speed} = 1500 \times 2 \times 10^5$$
$$= 1.5 \times 10^3 \times 2 \times 10^5$$
$$= 3 \times 10^{3+5} = 3 \times 10^8 \text{ m/s}$$

Brackets and indices

We see that

$$(x^3)^2 = x^3 \times x^3 = x^{3+3} = x^6$$

and

$$(y^2)^4 = y^2 \times y^2 \times y^2 \times y^2 = y^8$$

Rule When brackets are used, *multiply* the index in the bracket by the index outside.

Examples

$$(r^2)^6 = r^{12}; \qquad (x^3)^5 = x^{15}; \qquad (2y^4)^3 = 2^3 \times y^{12} = 8y^{12}$$

Division with indices

We have

$$x^5 \div x^2 = \frac{x^5}{x^2} = \frac{x \times x \times x \times x \times x}{x \times x} = x^3$$

We can simplify this by writing

$$\frac{x^5}{x^2} = x^{5-2} = x^3$$

Rule In division, *subtract* the indices.

Examples

1. $$\frac{a^6}{a^3} = a^{6-3} = a^3$$

2. $$\frac{(2y^2)^4}{(y^3)^2} = \frac{2^4 y^8}{y^6} = 16y^{8-6} = 16y^2$$

3. The force on an area of 2000 m² (2×10^3 m²) of a ship due to water pressure is 3×10^8 N (newton) (Figure 4.2). Calculate the average pressure.

$$\text{Average pressure} = \frac{\text{force}}{\text{area}}$$

$$= \frac{3 \times 10^8 \text{ N}}{2 \times 10^3 \text{ m}^2}$$

$$= 1.5 \times 10^{8-3} \text{ N/m}^2$$

$$= 1.5 \times 10^5 \text{ N/m}^2$$

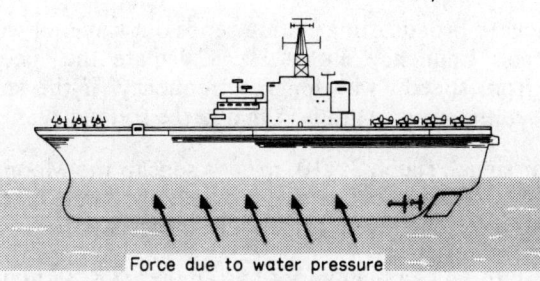

Figure 4.2.

EXERCISE 12

Simplify

1. $10^6 / 10^2$
2. $2r^2 \times 3r^3$
3. $(3y^2)^4 + (2y^4)^2$
4. $\frac{2x^8}{x^6}$

5. $\dfrac{3x^2}{x^5}$

6. $2(a^2)^6 + 3(a^4)^3$

7. $a^2 b^3$ if $a=2$ and $b=3$

8. $(x^2)^3 + \dfrac{x^4}{x}$ if $x=2$

9. $\dfrac{(a^4)^2}{(a^2)^3}$

10. $\dfrac{(y^2)^3 \times z^2}{y^2 \times (z^2)^4}$

11. Using powers of 10 in your answers, calculate:
    (i) 5000×200,
    (ii) $80\,000 \times 3000$,
    (iii) $1\,000\,000 \times 300$,
    (iv) $60\,000\,000 \div 20\,000$,
    (v) $400\,000 \div 200$.

12. Using the approximate formula $4 \times (\text{radius})^3$ for the volume of a sphere, calculate the volume in mm^3 of
    (i) a golf-ball of radius 20 mm,
    (ii) a cricket-ball of radius 35 mm,
    (iii) a football of radius 100 mm.
    Choose your answers in mm^3 from one of the values A to F:
    (A) 3.2×10^4,   (B) 1.715×10^5,   (C) 3.2×10^5,
    (D) 4×10^5,   (E) 1.715×10^6,   (F) 4×10^6.

13. What is the ratio of (i) the mass of the Sun, 2×10^{30} kg, to the mass of the Earth, 6×10^{24} kg, (ii) the mass of the Earth to the mass of the Moon, 7×10^{22} kg?

14. A long wave broadcasting station sends out waves of wavelength 10^3 m and frequency 3×10^5 Hz. Calculate the speed of the waves from speed = wavelength × frequency, if the speed is in m/s when the wavelength is in m and the frequency is in Hz.

15. A radar signal, speed 3×10^8 m/s, is sent to the Moon, which is 3.9×10^8 m from the Earth. Calculate the time taken by the signal to reach the Moon (time = distance/speed).

16. The base of an elephant's foot has an area 8×10^4 mm^2 and the force on it due to the elephant's weight is 12×10^3 N. Find the average pressure below the foot (average pressure = force/area).

17. There are about 6×10^{23} molecules in 2 g of hydrogen gas. Calculate the number of molecules in 4×10^3 g of hydrogen gas.

18. If 6×10^{23} molecules of air occupy a volume of about 2×10^4 cm^3, estimate the average number of molecules per cm^3.

Rules for directed numbers

Before dealing with negative indices it will be helpful to revise briefly positive and negative numbers and their rules. These are called *directed numbers* because the plus and minus can be linked respectively to opposite directions.

Since -2 and -3 can be considered two steps of 2 and 3 units in the same direction, we see that

$$-2 - 3 = -5$$

Similarly,

$$+1 + 2 = +3$$

Generally, then, we *add* the two numbers numerically when they have the same sign.

But $+2 - 3 = -1$ because $+2$ is a step of 2 units to the right say and -3 is a step of 3 units to the left, so the result is 1 step to the left. Similarly,

$$-3 + 1 = -2$$

and

$$+4 - 2 = +2$$

So we *subtract* the two numbers numerically when they have opposite signs and use the sign of the larger number with the answer.

Rules **1.** *Addition*   A *plus* sign does not affect the sign of a directed number. For example,

$$+(+2) = +2$$

and

$$+(-1) = -1$$

So

$$(-2) + (-3) = -2 - 3 = -5$$

and

$$(-2) + (+3) = -2 + 3 = +1$$

2. *Subtraction*   A *minus* sign changes the sign of the directed number. For example,

$$-(+2) = -2 \quad \text{and} \quad -(-1) = +1$$

So

$$(-2) - (+4) = -2 - 4 = -6$$

and

$$(-3) - (-5) = -3 + 5 = +2$$

3. Multiplication If the multiplied numbers have the *same* signs, the answer is +. For example,

$$(-2) \times (-3) = +6$$

and

$$(+4) \times (+2) = +8$$

If the multiplied numbers have *opposite* signs, the answer is −. For example,

$$(+4) \times (-1) = -4 \quad \text{and} \quad (-2) \times (+3) = -6$$

4. Division If the divided numbers have the *same* signs, the answer is +. For example,

$$\frac{-6}{-2} = +3 \quad \text{and} \quad \frac{+4}{+2} = +2$$

If the divided numbers have *opposite* signs, the answer is −. For example,

$$\frac{-4}{+2} = -2 \quad \text{and} \quad \frac{+3}{-1} = -3$$

Negative indices

We know that

$$10^3 = 1000, \quad 10^2 = 100 \quad \text{and} \quad 10^1 = 10$$

The next index after 3, 2, and 1, is 0, and since the numbers 1000, 100, 10 decrease 10 times when the index changes by 1, we expect that

$$10^0 = \frac{10^1}{10} = 1$$

We can prove $10^0 = 1$ from the rule for multiplication of numbers with indices. Thus

$$10^0 \times 10^2 = 10^{0+2} = 10^2$$

so

$$10^0 = \frac{10^2}{10^2} = 1$$

We now have $10^3 = 1000$, $10^2 = 100$, $10^1 = 10$, $10^0 = 1$. We can continue the index *below* 0 to −1, −2, and so on, by dividing by 10 each time. In this way we see that

$$10^0 = 1, \quad 10^{-1} = \frac{1}{10} = 0.1, \quad 10^{-2} = \frac{1}{100} = \frac{1}{10^2},$$

$$10^{-3} = \frac{1}{1000} = \frac{1}{10^3}$$

So

$$10^{-5} = \frac{1}{10^5} = \frac{1}{100\,000}, \quad 10^{-8} = \frac{1}{10^8} = \frac{1}{100\,000\,000}$$

We now see that

$$x^{-1} = \frac{1}{x^1} = \frac{1}{x}, \quad x^{-2} = \frac{1}{x^2}, \quad x^{-5} = \frac{1}{x^5}$$

So

$$4^{-2} = \frac{1}{4^2} = \frac{1}{16} \quad \text{and} \quad 5^{-3} = \frac{1}{5^3} = \frac{1}{125}$$

Multiplication As before, we *add* the indices. So, from our rules for directed numbers,

$$10^{-1} \times 10^{-3} = 10^{-4} = \frac{1}{10^4}$$

and

$$x^{-2} \times x^{-4} \times x^3 = x^{-2-4+3} = x^{-3} = \frac{1}{x^3}$$

Division As before, we *subtract* the indices. So, from our rules for directed numbers,

$$\frac{x^{-5}}{x^{-3}} = x^{-5-(-3)} = x^{-5+3} = x^{-2} = \frac{1}{x^2}$$

and

$$\frac{4 \times 10^{-3}}{2 \times 10^{-4}} = 2 \times 10^{-3-(-4)} = 2 \times 10^1 = 20$$

Examples

1. In the electrical resistance formula $R = \rho l / A$, calculate R numerically if $\rho = 2 \times 10^{-7}$, $l = 6$, $A = 3 \times 10^{-8}$.

We have

$$R = \frac{\rho l}{A} = \frac{2 \times 10^{-7} \times 6}{3 \times 10^{-8}}$$
$$= 4 \times 10^{-7-(-8)} = 4 \times 10^1$$
$$= 40$$

2. In an oil-film experiment for estimating the size d of an oil molecule, $d = V/A$, where V is the volume of oil and A is the area into which it spreads (Figure 4.3). Calculate d if $V = 5 \times 10^{-10}$ m^3 and $A = 10^{-3}$ m^2.

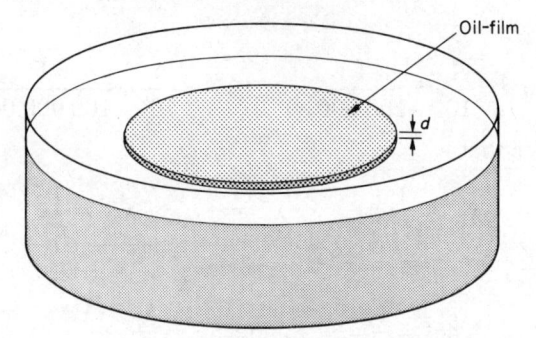

Figure 4.3.

We have $$d = \frac{V}{A} = \frac{5 \times 10^{-10}}{10^{-3}}$$
$$= 5 \times 10^{-10-(-3)} = 5 \times 10^{-7} \text{ m}$$

EXERCISE 13

Simplify the following, giving the numerical results in *fractions* in Questions 4 to 7:

1. (i) $-1+(-2)$ (ii) $-5-(-7)$ (iii) $(-2)\times(+3)$ (iv) $\dfrac{-4}{-2}$

2. (i) $-4-(-5)$ (ii) $+2-(-1)$ (iii) $\dfrac{6}{-2}$ (iv) $(-2)\times(-1)$

3. What is the *change* in temperature (i) from 20°C to −4°C, (ii) from −10°C to −8°C?

4. 2×10^{-2}

5. 4×10^{-1}

6. 3×10^{-3}

7. $\dfrac{a^{-2} \times a^{-3}}{4a}$

8. $\dfrac{4 \times 10^{-7} \times 3}{6 \times 10^{-8}}$

9. $\dfrac{6 \times 10^{-10}}{2 \times 10^{-4}}$

10. $\dfrac{2 \times 10^{-6} \times 3 \times 10^{-1}}{4 \times 10^{-8}}$

In the following questions, where possible give your answers with positive or negative indices.

11. Convert to m (metre): (i) 6 mm, (ii) 5 cm, (iii) 300 km, (iv) 5×10^{-5} cm.

12. Convert to m^2 (metre^2): (i) 5 mm^2, (ii) 6 cm^2, (iii) 20 mm^2, (iv) 10 km^2.

13. A sheet of paper has a thickness of 0.3 mm and a sheet of silver foil has a thickness of 2×10^{-4} mm. What are their thicknesses in metres (m)?

14. The frequency of a blue light is 6×10^{14} Hz (the Hz is a unit of frequency). Calculate the wavelength of the light wave if its speed is 3×10^8 m/s (wavelength = speed/frequency).

15. The wavelength of an orange light is 6×10^{-7} m and the speed of the light wave is 3×10^8 m/s. Calculate the frequency of the wave (frequency = speed/wavelength).

16. A radar signal is sent to an aeroplane 9×10^3 m away (Figure 4.4). If the speed of the signal is 3×10^8 m/s, find the time it takes to reach the aeroplane (time = distance/speed). How far away is another aeroplane if a radar signal takes 2×10^{-4} s to reach it?

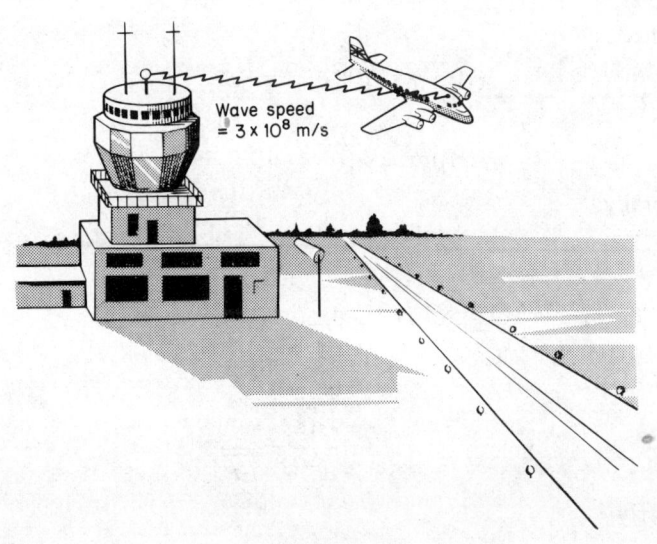

Figure 4.4.

17. The mass of silver deposited in electrolysis is 1.1×10^{-6} kg/C (kilogram per coulomb). Find the mass in kg deposited by (i) 8000 C, (ii) 6×10^5 C.

18. A sheet of paper has a length of 8 cm and a width of 12 cm. What is the area in m^2 after converting each length to metres?

19. A small cube has dimensions 8 mm × 8 mm × 8 mm. Convert the lengths of each side to metres and then calculate the volume of the cube in m^3.

Fractional indices

1. We now consider briefly fractional indices, for example, $10^{1/2}$. We know that
$$(10^{1/2})^2 = 10^1 = 10$$
So if $10^{1/2} = x$, then we can write
$$x^2 = 10. \quad \text{So } x = 10^{1/2} = \sqrt{10}$$
Similarly, since
$$(10^{1/3})^3 = 10^1 = 10, \quad \text{then}$$
$$10^{1/3} = \sqrt[3]{10}$$
Generally,
$$x^{1/3} = \sqrt[3]{x}$$

2. Since
$$(10^{2/3})^3 = 10^{(2/3)\times 3} = 10^2$$
then
$$10^{2/3} = \sqrt[3]{10^2}$$
Generally,
$$a^{2/3} = \sqrt[3]{a^2}$$

Negative fractions Since
$$a^{-2} = \frac{1}{a^2}, \quad \text{then}$$
$$a^{-2/3} = \frac{1}{a^{2/3}} = \frac{1}{\sqrt[3]{a^2}}$$
Similarly,
$$x^{-3/4} = \frac{1}{x^{3/4}} = \frac{1}{\sqrt[4]{x^3}}$$
and
$$y^{-2/5} = \frac{1}{y^{2/5}} = \frac{1}{\sqrt[5]{y^2}}$$
So
$$8^{-2/3} = \frac{1}{8^{2/3}} = \frac{1}{\sqrt[3]{8^2}} = \frac{1}{\sqrt[3]{64}} = \frac{1}{4}$$

EXERCISE 14

Calculate, or simplify, the following:

1. $4^{1/2}$ 2. $27^{1/3}$
3. $27^{2/3}$ 4. $100^{3/2}$
5. $1000^{2/3}$ 6. $(4a^2)^{1/2}$
7. $3(y^3)^{2/3}$ 8. $(16x^4)^{3/4}$
9. $100^{-1/2}$ 10. $100^{-3/2}$
11. $2y^{-3/2}$, where $y=4$ 12. $16^{-3/4}$

SOME AREAS AND VOLUMES

In planning rooms and buildings or laying gas pipes, for example, areas and volumes of rectangles, circles and cylinders may require calculation.

Areas

1. *Square or rectangle* The area of a square or rectangle = length × width. So a square of side 4 m has an area of

$$4 \text{ m} \times 4 \text{ m} = 16 \text{ m}^2$$

If a rectangle has sides 4 m by 2 m, then

$$\text{area} = 4 \text{ m} \times 2 \text{ m} = 8 \text{ m}^2$$

2. *Circle* The circumference of a circle $= 2\pi r$, where r is the radius (Figure 4.5(a)). The area of the circle $= \pi r^2$. So a rod of diameter 8 mm, or radius 4 mm, has a cross-sectional area (assuming $\pi = 3.14$) of

$$3.14 \times 4^2 = 50 \text{ mm}^2 \quad (2 \text{ s.f.})$$

3. *Cylinder* A length of pipe or wire is a cylinder (Figure 4.5(b)). If we 'unwind' a cylinder of length l and radius r, we see that the curved surface forms a rectangle of length l and width $2\pi r$, because $2\pi r$ is the circumference of the cross-section. So

$$\text{Area of curved surface of cylinder} = 2\pi r l$$

A length of 100 cm of pipe of radius 5 cm thus has a surface area of

$$2 \times 3.14 \times 5 \times 100 = 3140 \text{ cm}^2$$
$$= 0.314 \text{ m}^2$$

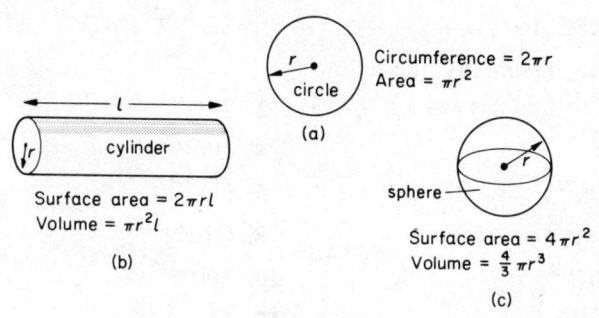

Figure 4.5.

4. Sphere The surface area of a sphere of radius $r = 4\pi r^2$ (Figure 4.5(c)). So a sphere of radius 10 cm has a surface area of

$$4 \times 3.14 \times 10^2 = 1256 \text{ cm}^2$$

In all area formulae, *two* lengths are multiplied, for example, $\pi r^2 = \pi \times r \times r$. So $2\pi r$ can not be an area; in fact, it is the *length* of the circumference of a circle.

Volumes

1. Cuboid If a room has the same cross-section area as we move up from the floor to the ceiling, which is usually the case, then we can say that

volume of room = cross-section area × height

$$= lbh$$

if l is the length, b is the width and h is the height. So if $l = 5$ m, $b = 4$ m, $h = 6$ m, then

$$\text{volume} = 5 \times 4 \times 6 = 120 \text{ m}^3$$

The volume is expressed in 'cubic metre' (m^3) since three lengths are multiplied.

2. Cylinder If a cylinder has radius r and length l, then

volume = cross-section area × length

$$= \pi r^2 l$$

A pipe of radius 5 cm and length 100 cm thus has a volume of

$$3.14 \times 5^2 \times 100 = 7850 \text{ cm}^3$$

3. Sphere If r is the radius of a sphere, then

$$\text{volume} = \tfrac{4}{3}\pi r^3$$

INDICES

So if the radius of a ball is 5 cm, its volume is:

$$\tfrac{4}{3} \times 3.14 \times 5^3 = 520 \text{ cm}^3 \quad (2 \text{ s.f.})$$

In all volume formulae, *three* lengths are multiplied, for example, $\pi r^2 l = \pi \times r \times r \times 1$ and $\tfrac{4}{3}\pi r^3 = \dfrac{4\pi}{3} \times r \times r \times r.$

Summary

	area	volume
Circle	πr^2	—
Cylinder	$2\pi r l$	$\pi r^2 l$
Sphere	$4\pi r^2$	$\tfrac{4}{3}\pi r^3$

Surface area to volume ratio in warm blooded animals

The heat lost by an animal to its surroundings is proportional to its *surface area*. The mass of the animal is proportional to its *volume*. So

$$\text{temperature drop per unit mass} \propto \frac{\text{surface area}}{\text{volume}}$$

We can see how the ratio *surface area/volume* depends on the size of the animal by considering the simple case of a cube. A cube of side 1 cm has six faces each of area 1 cm^2. So the surface area = 6 cm^2. Since the volume is 1 cm^3,

$$\frac{\text{surface area}}{\text{volume}} = \frac{6}{1} = 6$$

A larger cube of side 2 cm has six faces each of area $2 \times 2 = 4$ cm^2. So the total surface area = $6 \times 4 = 24$ cm^2. The volume of the cube = $2 \times 2 \times 2 = 8$ cm^3. So

$$\frac{\text{surface area}}{\text{volume}} = \frac{24}{8} = 3$$

A much larger cube such as one of side 20 cm has a surface area $6 \times 400 = 2400$ cm^2 and a volume $20 \times 20 \times 20 = 8000$ cm^3. In this case,

$$\frac{\text{surface area}}{\text{volume}} = \frac{2400}{8000} = 0.3$$

We now see from these results that the temperature drop of a large animal would be much less than that of a small animal when heat is lost to the surroundings. So a very large animal such as a whale in water has only a very small temperature drop. These animals are *warm-blooded*; they can keep their temperature well above that of the water in which they live. On the other hand, small sea-creatures suffer an appreciable temperature drop in cold water. These creatures are *cold-blooded*, that is, their body temperature changes with the temperature of the surroundings.

EXERCISE 15

(Where necessary, assume $\pi = 3.14$.)

1. A circular disc has a radius of 10 mm and a thickness of 2 mm. Find (i) its circumference, (ii) the area of one face, (iii) the volume of the disc.

2. A wire X has twice the radius and twice the length of a wire Y. What is the ratio of (i) the circumferences of their cross-sections, (ii) their cross-section areas, (iii) their volumes?

3. Calculate the cross-section area of a wire of diameter 0.6 mm (2 s.f.).

4. Find the surface area and volume of a ball-bearing of diameter 4 mm (2 s.f.).

5. A ball X has half the radius of another ball Y. What is the ratio of (i) their surface areas, (ii) their volumes?

6. A pipe has an average radius of 2 cm and is 80 cm long. Find (i) the curved surface area, (ii) the volume occupied by the pipe.

7. The surface area of a sphere X is four times that of a sphere Y. Calculate the ratio of (i) their radii, (ii) their volumes.

8. A box is a cube with each side 21 cm long. How many tennis-balls, each of diameter 7 cm, can be packed tightly in the box?
   What is the volume of the air space in the box, to the nearest 100 cm^3, after all the balls are packed into it?

9. Two hot metal spheres, at the same temperature, are suspended in a room. One has a radius of 2 cm and the other a radius of 20 cm. What is the ratio of their temperature fall per second? (Heat lost $\propto$ surface area; mass $\propto$ volume.)

5 Logarithms

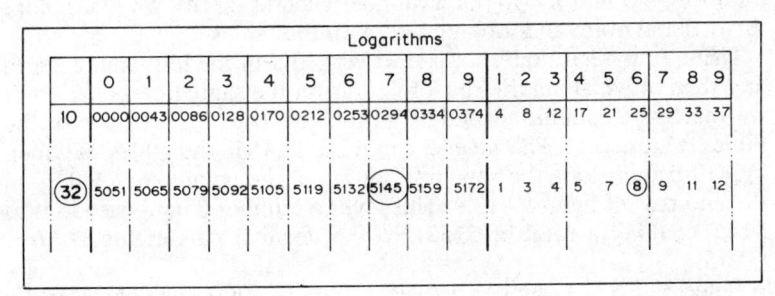

	Logarithms																		
	0	1	2	3	4	5	6	7	8	9	1	2	3	4	5	6	7	8	9
10	0000	0043	0086	0128	0170	0212	0253	0294	0334	0374	4	8	12	17	21	25	29	33	37
(32)	5051	5065	5079	5092	5105	5119	5132	(5145)	5159	5172	1	3	4	5	7	(8)	9	11	12

Figure 5.1.

1. Logarithms ('logs') are *indices* or *powers*. We know that
$$10 = 10^1, \quad 100 = 10^2, \quad 1000 = 10^3,$$
We say
   1 is the log of 10 to the base 10
   2 is the log of 100 to the base 10
   3 is the log of 1000 to the base 10

Logs may be indices or powers of numbers other than 10 but we shall use only the base 10. So log 1000 = 3 from above.

2. Since log 10 = 1 and log 100 = 2, it follows that the logs of numbers between 10 and 100 are between 1 and 2. For example, from Logarithm Tables, we know that
$$\log 20 = 1.301, \quad \text{and} \quad \log 35 = 1.489$$
This means that $10^{1.301} = 20$ and $10^{1.489} = 35$.

Similarly, numbers between 100 and 1000 have logs greater than 2 and less than 3. For example,
$$\log 300 = 2.4771, \quad \log 800 = 2.9030$$

3. The whole number used in the log is called its *characteristic* (it tells you whether the number is in tens or hundreds or thousands or other multiples of 10). The decimal part is called its *mantissa*. So log 7.4, a number between 1 and 10, has a characteristic 0; log 63.2, a number between 10 and 100, has a characteristic 1; and so on.

The characteristic of a large number such as 87 325 may be obtained by counting 0, 1, 2, 3 ... from left to right. In this case we arrive at the number 4. As a check, $10^4 = 10\,000$, so 87 325, which is more than 10 000 and less than 100 000, has a characteristic 4.

Logarithm tables

Log tables provide the *decimal* part (mantissa) of the logarithm of numbers. To find the log of a number such as 327.6, we first ignore the decimal point and look up '3276' in the tables.

Using four-figure tables, we start with '32' in the left hand column and then move along the row of figures on the same line as '32' until we reach the column headed '7' in the row 0 to 9 at the top of the tables (Figure 5.1). This gives the number 5145 in the tables. We then move further along the row until we reach the column headed '6' in the end row of figures 1 to 9. This gives a number 8, and we add it to 5145, making a total of 5153. So the decimal part of log 327.6 is .5153.

Since 327.6 is a number between 100 and 1000, the characteristic or whole number part of the log is 2, from above. Hence

$$\log 327.6 = 2.5153$$

Summary To find the log of a number, (i) ignore the decimal point if this is present and use the tables to find the decimal part of the log, then (ii) add the whole number part by using 0 for units, 1 for tens, 2 for hundreds, and so on.

Check these logs from four-figure tables:

$\log 3.142 = 0.4972$ $\log 5555 = 3.7447$

$\log 60.75 = 1.7836$ $\log 981.3 = 2.9918$

Antilogs

If the log of a number is 2.3864, for example, we can use *antilogarithm* tables to find the number.

We first ignore the whole number or characteristic 2, and look up the first two figures '38' of the decimal in the left hand column in the tables (Figure 5.2). We then move along this row until we reach '6' in the top row of figures 0 to 9. This gives the number 2432. We now move further along the row until we reach the column headed '4' in the top row of figures 1 to 9. This gives a number 2, and adding it to 2432 we obtain a total 2434.

LOGARITHMS

Anti-logarithms

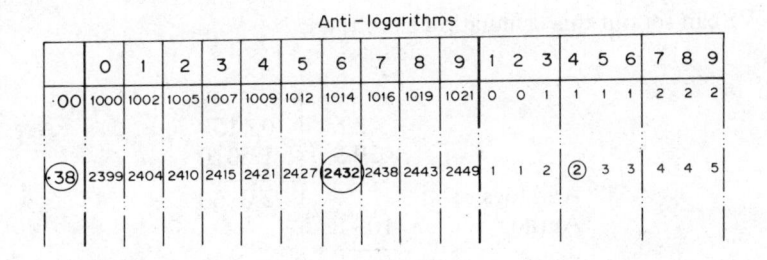

	0	1	2	3	4	5	6	7	8	9	1	2	3	4	5	6	7	8	9
·00	1000	1002	1005	1007	1009	1012	1014	1016	1019	1021	0	0	1	1	1	1	2	2	2
(38)	2399	2404	2410	2415	2421	2427	(2432)	2438	2443	2449	1	1	2	(2)	3	3	4	4	5

Figure 5.2.

So log 2434 is a number whose decimal part is .3864. Since the log of the number we need is 2.3864, the number must be in hundreds as the characteristic is 2. So the number if 243.4, placing the decimal point as here.

If the antilog of 5.3864 is needed, then the number is between 100 000 and 1 000 000. So we add noughts after 2434, giving the number 243 400.

Summary (i) Look up the decimal part of the log in the antilog tables. (ii) From the whole number or characteristic of the log, place a decimal point in the required position or add the number of noughts required.

EXERCISE 16

Using four-figure log tables, find the logs of

1. 63.72 **2.** 1.483 **3.** 745.6
4. 31.59 **5.** 83 120

Using four-figure log tables, find the antilogs of

6. 2.6550 **7.** 0.4972 **8.** 1.8745
9. 3.0771 **10.** 6.6782

Multiplication and division using logs

1. We know from the section on Indices that we *add* indices when *multiplying* two quantities. Now a log is an index or power. So when we multiply two numbers, we *add their logs*. For example,

$$2.6 \times 39.9 = 10^{0.4150} \times 10^{1.6010} = 10^{2.0160}$$

When we look up in tables the number whose antilog is 2.0160, we obtain 103.8. So

$$2.6 \times 39.9 = 103.8 \text{ to 4 s.f.}$$

We can set out this calculation as follows:

	No.	Log
	2.6	0.4150
	×39.9	1.6010
Add logs		2.0160
Antilog	103.8	

2. When we *divide* two numbers expressed with indices, we *subtract* the indices (p. 31). So we subtract the logs when two numbers are divided. For example,

$$\frac{39.9}{2.6} = \frac{10^{1.6010}}{10^{0.4150}} = 10^{1.6010-0.4150} = 10^{1.1860}$$

The number whose antilog is 1.1860 is 15.35. So

$$\frac{39.9}{2.6} = 15.35 \text{ to 4 s.f.}$$

We can set out this calculation as follows:

	No.	Log
	39.9	1.6010
	÷ 2.6	0.4150
Subtract logs		1.1860
Antilog	15.35	

EXERCISE 17

Using logs, find the value of the following to 1 decimal place.

1. 7.352×18.94
2. 18.6×30.87
3. $46.72 \div 8.54$
4. $8362 \div 346.7$
5. $\dfrac{256.3 \times 7.492}{89.96}$

Logs of decimals or fractions

We now consider logs of numbers *less* than 1, for example 0.37 or 0.042 or 0.0086.

We can write

$$0.37 = \tfrac{1}{10} \times 3.7 = 10^{-1} \times 3.7$$

LOGARITHMS

Since we *add* the logs of two numbers multiplied together, then

$$\log 0.37 = \log 10^{-1} + \log 3.7$$

Now $\log 10^{-1} = -1$, since -1 is the power of 10, and, from tables, $\log 3.7 = .5682$. So

$$\log 0.37 = -1 + .5682.$$

For short, we write this

$$\log 0.37 = \bar{1}.5682$$

The minus is placed above the 1 to save space and the plus sign in front of the decimal is omitted for the same reason. Remember that *the decimal part of a log is always positive.* So in

$$\log 0.51 = \bar{1}.7076$$

the log is actually $-1 + .7076$ or $-1 + 0.7076$.

We now see that

$$\log 0.1 = \log \tfrac{1}{10} = \log 10^{-1} = -1 \text{ or } \bar{1}$$
$$\log 0.01 = \log 10^{-2} = -2 \text{ or } \bar{2}$$
$$\log 0.001 = -3 \text{ or } \bar{3}$$

and so on. The whole number following the minus sign is *one more* than the number of noughts in the decimal. So $\log 0.000\,001 = -6$ or $\bar{6}$.

From log tables, check the following:

$\log 0.4 = \bar{1}.6021, \quad \log 0.068 = \bar{2}.8325, \quad \log 0.00824 = \bar{3}.9159$

Multiplying decimals using logs

Suppose we need to find 0.07462×0.389. As on p. 46, we can set out this multiplication sum as follows.

No.	Log
0.07462	$\bar{2}.8728$
×0.389	$\bar{1}.5899$

We now have to add the logs, remembering that *the decimal parts are positive.* So their total is $+1.4627$. We therefore carry $+1$ to the whole number column, which has negative numbers -2 and -1, making a final total of -2 or $\bar{2}$. So the result of adding the two logs is $\bar{2}.4627$. The antilog is 0.02902. So

$$0.07462 \times 0.389 = 0.02902 \quad \text{to 4 s.f.}$$

Dividing decimals using logs

1. Suppose we now need the answer to the division 0.00654/0.369. As before,

No.	Log
0.00654	$\bar{3}.8156$
÷ 0.369	$\bar{1}.5670$

In division, we *subtract* the two logs. The decimal parts are positive and their difference is .2486. The characteristics are both negative and their difference is

$$\bar{3} - \bar{1} = (-3) - (-1) = -3 + 1 = -2 \text{ or } \bar{2}$$

So subtraction of the logs gives an answer $\bar{2}.2486$, whose antilog is 0.01813 or 0.018 to 2 s.f.

You should always *check subtraction* of two logs by adding your result to the lower line. In this case, for example, $\bar{2}.2486 + \bar{1}.5670 = \bar{3}.8156$, which is the same as the top log above.

2. A more awkward case of subtracting two logs is shown below.

Log

$$\begin{array}{r} {\scriptstyle -1+1} \\ \bar{1}.3860 \\ -\bar{2}.4940 \end{array}$$

Here we 'borrow' +1 to make the subtraction of the decimal parts positive, giving $1.3860 - .4940 = .8920$, and compensate by adding −1 to $\bar{1}$, which gives $-1 - 1 = -2$ or $\bar{2}$ at the top. Subtraction of the lower $\bar{2}$ now gives the answer 0. So the final answer for the subtraction is

0.8920

As a check, add 0.8920 to the lower line $\bar{2}.4940$. The result is $\bar{1}.3860$, which is the top log. The antilog of 0.8920 is 7.798 or 7.80 to 3 s.f.

EXERCISE 18

Calculate the following to 2 s.f. in each case.

1. 0.326×0.065 **2.** $\dfrac{0.862}{0.034}$ **3.** 0.00891×0.75

4. $\dfrac{0.00615}{0.0873}$ **5.** $\dfrac{0.946 \times 0.0175}{0.0036}$

Powers using logs

To calculate 0.036^2, we have to multiply the log of 0.036 by 2. So

No.	Log
0.036	$\bar{2}.5563$
0.036^2	$\times 2$
	$\bar{3}.1126$

$0.5563 \times 2 = 1.1126$; the 1 is added to 2×-2 or -4, making -3 or $\bar{3}$.
The antilog of $\bar{3}.1126$ is 0.001296, so the result is 0.0013 to 2 s.f.
As another example, consider 0.72^3. Then

No.	Log
0.72	$\bar{1}.8573$
0.72^3	$\times 3$
	$\bar{1}.5719$

The antilog of $\bar{1}.5719$ is 0.3732 or 0.37 to 2 s.f.

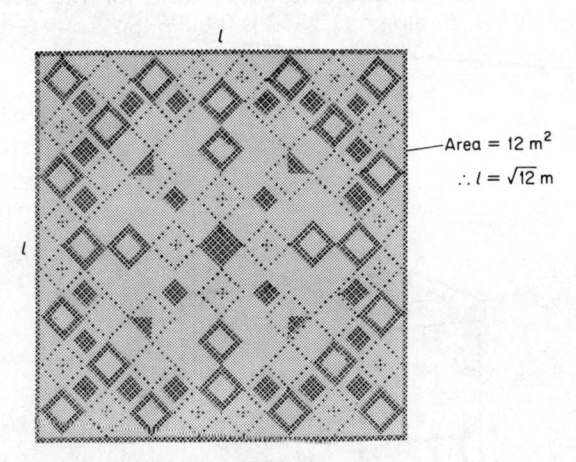

Figure 5.3.

Roots using logs

1. Suppose we require $\sqrt{12}$, the square root of 12 (Figure 5.3). Now $\sqrt{12} = 12^{1/2}$. So we need to divide the log by 2. We set this out as follows.

BASIC MATHEMATICS FOR SCIENCE

No.	Log
12	1.0792
$\sqrt{12}$	0.5396

We have divided the log 1.0792 by 2. The antilog of 0.5396 is 3.463. So

$$\sqrt{12} = 3.46 \quad \text{to 3 s.f.}$$

2. Suppose we require $\sqrt{0.372}$. As before,

No.	Log
0.372	$\bar{1}.5705$
$\sqrt{0.372}$	$\bar{1}.7852$

Dividing $\bar{1}.5705$ by 2 is more awkward than in the previous example above. We need a negative *whole number* in the characteristic, so we subtract 1 from $\bar{1}$, giving -2, and compensate by adding 1 to the decimal part, giving 1.5705. When we now divide by 2, the result is -1 or $\bar{1}$ for the characteristic, and .7852 for the decimal part ($\bar{1}.5705 \div 2$). The antilog of $\bar{1}.7852$ is 0.6098. So

$$\sqrt{0.372} = 0.61 \quad \text{to 2 s.f.}$$

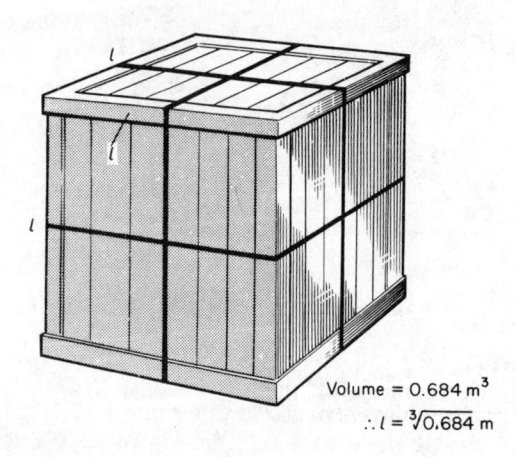

Volume = 0.684 m³

∴ $l = \sqrt[3]{0.684}$ m

Figure 5.4.

LOGARITHMS

3. To find $\sqrt[3]{0.648}$ (Figure 5.4), we must divide the log by 3. So

No.	Log
0.648	$\bar{1}.8116$
$\sqrt[3]{0.648}$	$\bar{1}.9372$

Here we subtract 2 from $\bar{1}$, giving -3, and compensate by adding 2, making the decimal part 2.8116. On dividing by 3, the characteristic is -1 or $\bar{1}$ and the decimal part is 0.9372. The antilog of $\bar{1}.9372$ is 0.8654. So

$$\sqrt[3]{0.648} = 0.87 \quad \text{to 2 s.f.}$$

EXERCISE 19

Calculate the following to 2 s.f.

1. 0.086^3

2. $\sqrt{0.6982}$

3. $\sqrt[3]{0.926}$

4. $\dfrac{0.064^2 \times 3.142}{4}$

5. $\dfrac{\sqrt{0.43} \times 0.0507}{0.27^3}$

ALGEBRA

6 Linear Equations

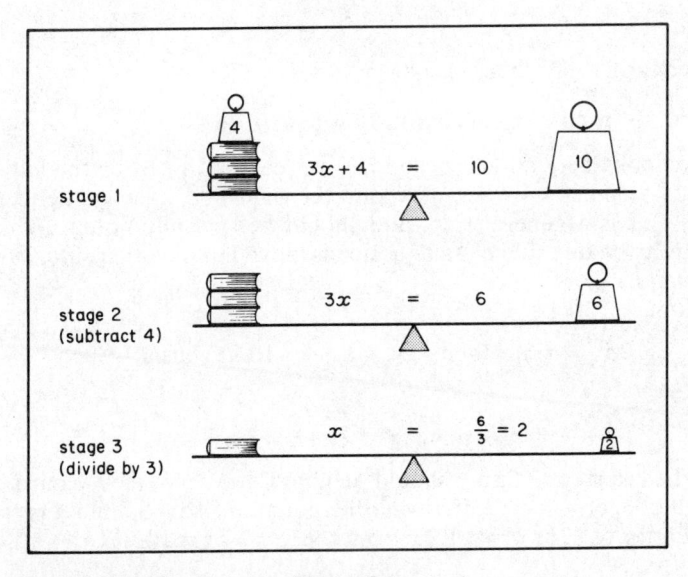

Figure 6.1.

A *linear* equation in x is an equation which contains x to the power of 1. So examples of a linear equation are

$$3x + 4 = 10 \quad \text{or} \quad 7 - 5x = 2$$

We often obtain linear equations in simple problems. In Figure 6.1, for example, 3 books and a weight of 4 units together balance a weight of 10 units. So if each book has a weight represented by x units, we can say that

$$3x + 4 = 10$$

In any equation, linear or otherwise, we can make any change to one side of an equation provided we make the *same* change to the other side. Otherwise the balance is upset. To find x, we can proceed as follows.

$$3x + 4 = 10 \qquad (1)$$

(subtract 4 from each side)

$$3x + 4 - 4 = 10 - 4$$
$$3x = 10 - 4 \qquad (2)$$

So
$$3x = 6$$

(divide each side by 3)

$$x = \frac{6}{3} = 2$$

Check If $x = 2$, then $3x + 4 = 6 + 4 = 10$

How to 'move' quantities in equations

If we look at equations (1) and (2), we see that the +4 on the left side of (1) becomes −4 when moved to the right side, as seen in equation (2). This is a general rule, and should be specially noted: When a quantity on one side of an equation is moved to the other side, *its sign changes*.

So if
$$2a - 4 = 6, \quad \text{then} \quad 2a = 6 + 4 = 10. \quad \text{Thus } a = \frac{10}{2} = 5$$

and if
$$6 - y = 2y, \quad \text{then} \quad 6 = 2y + y = 3y. \quad \text{So } y = \frac{6}{3} = 2$$

The last step to find y should be noted. If $3y = 6$, then y must be a third of 6, or $y = \frac{6}{3} = 2$. If, in another equation, $4a = 3$, then a must be a quarter of 3, or $a = \frac{3}{4}$. If $\frac{1}{2}x = 5$, then $x = 2 \times 5 = 10$. If

$$\tfrac{2}{3}x = 4$$

then
$$2x = 4 \times 3 = 12$$

So
$$x = \frac{12}{2} = 6$$

Example

Solve the equation $4y - 3 = y + 6$.

With linear equations, we always collect the unknown quantities (those with y) on one side of the equation, and the known quantities (the numbers) on the other side of the equation. On moving the quantities, remember that 'y' means '$+y$' so that the change in sign is to '$-y$'. On moving y in the above equation to the left side, and −3 to the right side, we obtain

$$4y - y = +6 + 3$$

LINEAR EQUATIONS

So
$$3y = 9$$
and
$$y = \frac{9}{3} = 3$$

Check Left side of original equation $= 4y - 3 = 12 - 3 = 9$, when $y = 3$.
Right side of original equation $= y + 6 = 3 + 6 = 9$, when $y = 3$.

Linear equations with brackets

Here we first remove the brackets and then proceed as before. For example, suppose we have to find the unknown value of E from

$$4(2E - 4) + 7 = 3(E - 1) + 9$$

Remove brackets,
$$8E - 16 + 7 = 3E - 3 + 9$$

Collect unknowns and knowns,
$$8E - 3E = -3 + 9 + 16 - 7$$
$$\therefore \ 5E = 15$$
$$E = \frac{15}{5} = 3$$

Check Left side of original equation $= 4(2E - 4) + 7 = 4 \times 2 + 7 = 15$.
Right side of original equation $= 3(E - 1) + 9 = 3 \times 2 + 9 = 15$.

EXERCISE 20

Solve the following equations and check your answers.

1. $2a - 6 = 4$
2. $10 = 3E - 1$
3. $3y - 2 = 8 - 2y$
4. $6I + 5 = 13 - 2I$
5. $2(r + 6) = 18$
6. $24 = 4(x - 6) + 12$
7. $3(x + 5) = 5(x - 1)$
8. $2a - 6 = 6a - 6$
9. $3(4 - 2I) + 4 = 2(4 - I)$
10. $4(x - 6) - 6 = 2(5 - 2x)$

Linear equations with fractions

In equations with fractions, start by multiplying both sides by the number or quantity which will get rid of *all* the denominators. This number or quantity should be the lowest into which all the denominators will divide, or 'lowest common multiple' (l.c.m.).

As examples, consider the following fractional equations.

1. Solve
$$\frac{3x}{4} - \frac{1}{6} = \frac{x}{3} + \frac{2}{3}$$

Here the denominators 4, 6, 3, 3 will all divide into 12, which is the lowest common multiple. So multiply *both* sides by 12.
Now

$$\frac{3x}{4} \times 12 = 9x, \quad \frac{1}{6} \times 12 = 2, \quad \frac{x}{3} \times 12 = 4x, \quad \text{and} \quad \frac{2}{3} \times 12 = 8$$

So we have

$$9x - 2 = 4x + 8$$
$$\therefore \quad 9x - 4x = +8 + 2$$
$$5x = 10$$
$$x = \frac{10}{5} = 2$$

2. Solve

$$\frac{1}{y} + \frac{7}{8} = \frac{2}{3} + \frac{3}{2y}$$

The denominators y, 8, 3, and $2y$ will all divide into $24y$, the lowest common multiple. So multiply both sides by $24y$. The result is

$$24 + 21y = 16y + 36$$
$$\therefore \quad 21y - 16y = 36 - 24$$
$$5y = 12$$
$$y = \frac{12}{5} = 2.4$$

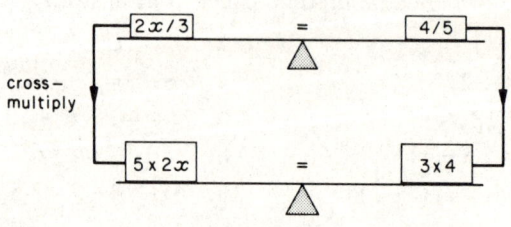

Figure 6.2.

Cross-multiplying in fractional equations

Suppose we have to solve a simple fractional equation such as

$$\frac{2x}{3} = \frac{4}{5}$$

If we multiply both sides by 15, the result is

$$5 \times 2x = 3 \times 4$$

LINEAR EQUATIONS

So in the fractional equation of this type, the 5 in the denominator on the right side can be brought across to the other side to multiply the numerator $2x$; and the denominator 3 on the left side can be brought across to the right side to multiply the numerator 4. We call this procedure 'cross-multiplying' and it is useful for fractional equations of this type *but no other* (see below).

We now proceed to solve the linear equation $5 \times 2x = 3 \times 4$, as the following examples illustrate.

Examples

1. Solve $\dfrac{3y}{5} = \dfrac{9}{10}$.

Cross-multiply,
$$10 \times 3y = 5 \times 9$$
$$30y = 45$$
$$y = \frac{45}{30} = \frac{3}{2} = 1\tfrac{1}{2}$$

2. Solve $\dfrac{3}{8R} = \dfrac{1}{2}$.

Cross-multiply,
$$2 \times 3 = 8R \times 1$$
$$6 = 8R$$
$$R = \frac{6}{8} = \frac{3}{4}$$

3. Solve $\dfrac{2(x-4)}{3} = \dfrac{2x-5}{4}$.

Cross-multiply,
$$4 \times 2(x-4) = 3 \times (2x-5)$$
$$8x - 32 = 6x - 15$$
$$8x - 6x = -15 + 32$$
$$2x = 17,$$
$$x = \frac{17}{2} = 8\tfrac{1}{2}$$

Limitations of cross-multiplying

It should be carefully noted by the student that 'cross-multiplication' can only be used when an equation is of the fractional type $a/b = c/d$, that is, the equation has *one* fraction only on *both* sides of the equal sign.

An equation such as

$$\frac{x}{4}+\frac{2}{5}=\frac{2x}{3}$$

has *two* fractions added together on one side of the equals, so cross-multiplying cannot be used. In this case we follow the procedure described on p. 57, that is, we multiply throughout by the l.c.m. of 4, 5, and 3, which is 60, to clear all the fractions.

Some simple fractional equations

Simple fractional equations are met in optical formulae or in electric circuit formulae.

As an example consider the equation

$$\frac{1}{x}+\frac{1}{30}=\frac{1}{10}$$

As in all linear equations, take all the known fractions to one side. So

$$\frac{1}{x}=\frac{1}{10}-\frac{1}{30}$$

$$\frac{1}{x}=\frac{3}{30}-\frac{1}{30}=\frac{2}{30}$$

Cross-multiply,

$$30 \times 1 = x \times 2$$

$$30 = 2x$$

$$x = \tfrac{30}{2} = 15$$

EXERCISE 21

Solve the following equations:

1. $\dfrac{2x}{3}-\dfrac{1}{2}=\dfrac{x}{6}+\dfrac{2}{3}$

2. $\dfrac{1}{v}-\dfrac{1}{12}=\dfrac{1}{4}$

3. $\dfrac{1}{R}=\dfrac{1}{3}+\dfrac{1}{6}$

4. $\dfrac{2}{s}=\dfrac{7}{3}$

5. $\dfrac{y-2}{3}-\dfrac{1}{2}=\dfrac{y}{6}+\dfrac{3}{4}$

6. $\tfrac{1}{2}(a-4)+\tfrac{1}{3}=\tfrac{1}{3}(a-1)+\tfrac{1}{2}$

7. $\dfrac{4}{3R}=\dfrac{1}{6}$

8. $\dfrac{3}{5u}=\dfrac{2}{3}$

9. $\dfrac{4}{3u}-\dfrac{1}{2}=\dfrac{1}{u}+\dfrac{1}{4}$

10. $\dfrac{2(x-4)}{3}-\dfrac{x}{6}=2+\dfrac{x-1}{6}$

7 Applications of Formulae in Physics

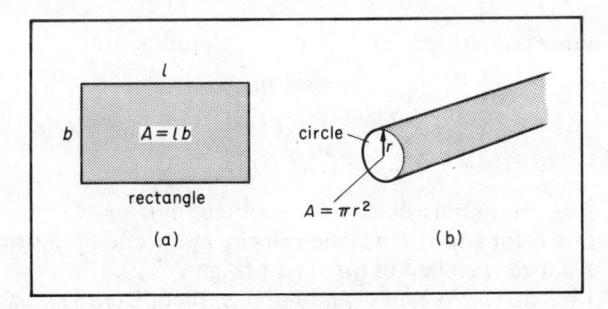

Figure 7.1.

An algebraic *formula* is a shorthand way of describing the relationship between quantities. For example,

$$A = lb$$

states that the area A of a rectangle is equal to its length l multiplied by its breadth b (Figure 7.1(a)). So if $l = 20$ m and $b = 15$ m, then

$$A = 20 \times 15 = 300 \text{ m}^2$$

The area A of a circle of radius r is given by (Figure 7.1(b)):

$$A = \pi r^2$$

So if the radius r of a wire is 2.0 mm, the area A of the circular cross-section, using $\pi = 3.14$, is given by

$$A = \pi \times 2.0^2 = 3.14 \times 2.0^2$$

$$= 12.56 = 12.6 \text{ mm}^2 \quad (1 \text{ dec. place})$$

MECHANICS

Linear motion with uniform acceleration

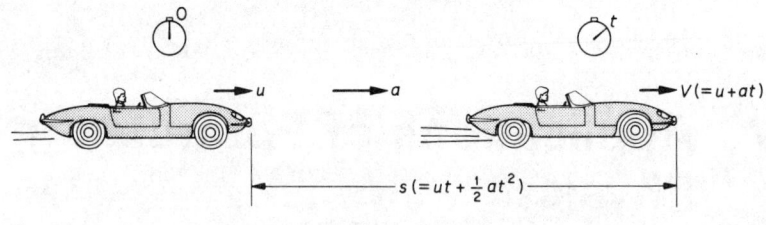

Figure 7.2.

The equations of motion for uniform acceleration are

$$v = u + at \qquad (1)$$
$$s = ut + \tfrac{1}{2}at^2 \qquad (2)$$
$$v^2 = u^2 + 2as \qquad (3)$$

Here u is the initial velocity of an object moving with a constant acceleration a for a time t, v is the velocity at the end of this time, and s is the distance travelled in the time t (Figure 7.2).

So if $u = 8$ m/s, $a = 2$ m/s^2 and $t = 10$ s, then, from (1), final velocity $v = u + at = 8 + (2 \times 10) = 28$ m/s. From (2),

$$\text{distance } s = ut + \tfrac{1}{2}at^2$$
$$= (8 \times 10) + (\tfrac{1}{2} \times 2 \times 10^2)$$
$$= 80 + 100 = 180 \text{ m}$$

Suppose an object has a *deceleration*, that is, a negative acceleration, of 10 m/s^2. Then $a = -10$ m/s^2. If the object had an initial velocity $u = 20$ m/s, and travelled a distance $s = 15$ m under this deceleration, the final velocity v is given, from (3) above, by

$$v^2 = u^2 + 2as$$
$$= 20^2 + (2 \times -10 \times 15) = 400 - 300$$
$$= 100$$
$$\therefore \quad v = \sqrt{100} = 10 \text{ m/s}$$

EXERCISE 22

Equations of motion (where necessary, assume $g = 10$ m/s^2).

1. A car moving at 15 m/s (u) accelerates at 3 m/s^2 (a) for 10 s (t). Find (i) the final velocity (v), (ii) the distance (s) travelled during acceleration.

APPLICATIONS OF FORMULAE IN PHYSICS 63

2. A train starts from rest and accelerates at 4 m/s^2 for 10 s. Find (i) the final velocity, (ii) the distance travelled during acceleration.

3. A car moving at 20 m/s brakes (decelerates) at 2 m/s^2 for a distance of 36 m. Find the final velocity of the car.

4. A ball is thrown vertically downward with a velocity of 10 m/s from a window at a height of 15 m. Calculate the velocity of the ball just before it hits the ground (assume g = acceleration due to gravity = 10 m/s^2).

5. A ball is thrown vertically upwards from the ground with an initial velocity of 20 m/s. (i) How long does it take to reach its highest point? (ii) How high does it travel? (Assume deceleration = $-g$ = -10 m/s^2.)

Forces; weight

In SI units, the newton, symbol N, is the unit of force. It is the force which gives a mass of 1 kg an acceleration of 1 m/s^2.

A force F which gives a mass m an acceleration a is given by (Figure 7.3(a))

$$F = ma$$

So if m = 1200 kg and a = 3 m/s^2, the force F is given by

$$F = ma = 1200 \times 3 = 3600 \text{ N}$$

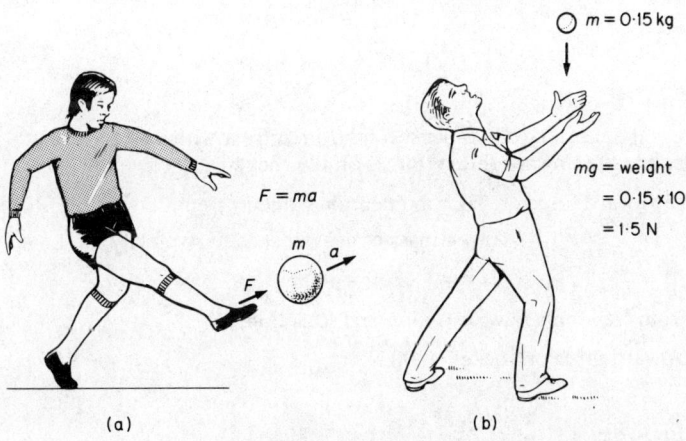

Figure 7.3.

The weight of an object is the force on it due to the downward gravitational pull of the earth. If the object falls freely, its acceleration would be about 10 m/s^2. Suppose the object has a mass

50 kg. Then, from $F = ma$,

$$\text{weight} = ma = 50 \times 10 = 500 \text{ N}$$

The acceleration of free fall is given the symbol g. So (Figure 7.3(b))

$$\text{weight} = mg$$

At the Moon's surface, the acceleration of free fall is only about 1.6 m/s^2. So a mass of 50 kg there would only have a weight of

$$50 \times 1.6 = 80 \text{ N}$$

Momentum and force

The linear *momentum* of an object of mass m moving with a velocity u is defined as the product $m \times u$ or mu.

So when $m = 60$ kg and $u = 10$ m/s,

$$\text{momentum} = 60 \times 10 = 600 \text{ kg m/s}$$

If the velocity is suddenly increased by a force to 15 m/s in the same direction, the new momentum is $60 \times 15 = 900$ kg m/s. So

$$\text{momentum change} = 900 - 600 = 300 \text{ kg m/s}$$

Now Newton defined force as the 'momentum change per second' it produces. So if the pushing force acts for 2 s,

$$\text{force} = \frac{\text{momentum change}}{\text{time}}$$

$$= \frac{300}{2} = 150 \text{ N}$$

Example

A rocket emits 3 kg of gas per second from its rear with a velocity of 200 m/s. Calculate the upward thrust (force) on the rocket.

Downward force on gas = momentum gained by emitted gas per second

$$= \text{mass of gas per second} \times \text{velocity}$$

$$= 3 \times 200 = 600 \text{ N}$$

From Newton's Law of Action and Reaction,

Upward force on rocket = 600 N

EXERCISE 23

Force; *weight*; *momentum* (where necessary, assume $g = 10 \text{ m/s}^2$).

1. Find the force which gives a ball of mass 0.2 kg an acceleration of 5 m/s^2. What is the weight of the ball?

APPLICATIONS OF FORMULAE IN PHYSICS 65

2. Calculate the force which gives an acceleration of 3 m/s^2 to a car of mass 1000 kg. What is the weight of the car?

3. A boy of mass 40 kg, moving with a velocity of 2 m/s, is pushed so that he starts to move with a velocity of 8 m/s in the same direction. What is the momentum change of the boy? What is the pushing force if it acts for 3 s?

4. A ball of mass 0.1 kg is thrown vertically upwards. Its initial velocity is 20 m/s and after 1 s this decreases steadily to 10 m/s. (i) What is the momentum change of the ball? (ii) What force has produced this change?

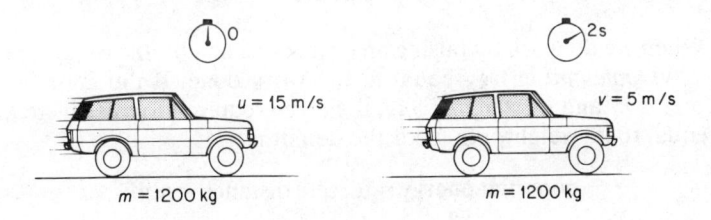

Figure 7.4.

5. A car of mass 1200 kg is travelling at 15 m/s and this velocity decreases steadily to 5 m/s when brakes are applied for 2 s (see Figure 7.4). (i) What is the momentum change of the car? (ii) What is the deceleration? (iii) What force has produced it?

6. Calculate the upthrust or force on a rocket which emits 0.5 kg of gas from its rear with an average velocity of 100 m/s.

7. A man of mass 60 kg, travelling in a car at 54 km/h (15 m/s) with a seat belt round him, has to brake in an emergency. If the belt brings him to rest in 0.5 s, what average force was exerted by the belt on the man?

Work and potential energy

The *joule*, symbol J, is the unit of work or energy in SI units. It is defined as the work done when a force of 1 N moves through 1 m in the direction of the force. Roughly, 1 J is the work done in raising an average-sized apple from the floor to the table.

Generally, *work* is defined as 'force × distance moved in direction of force'. So if a sledge is pulled by a constant force of 100 N through a distance of 20 m in the direction of the force,

$$W = 100 \text{ N} \times 20 \text{ m} = 2000 \text{ J}$$

Figure 7.5.

When we do work by raising an object to a height, the object is said to have *potential energy* equal to the work done. If the mass of the object is m and its weight is mg, the force required to raise it steadily is equal to its weight, mg. So if the height is h,

$$\text{potential energy} = \text{force} \times \text{distance} = mgh$$

The potential energy is in joules if m is in kg, g is about 10 m/s^2 and h is in m.

Example

Calculate the potential energy relative to its initial position of (i) a rocket of mass 1000 kg when it is moving at a height of 50 m above the ground, (ii) a man of mass 60 kg when on a diving board 4 m above the pool as in Figure 7.6 ($g = 10 \text{ m/s}$).

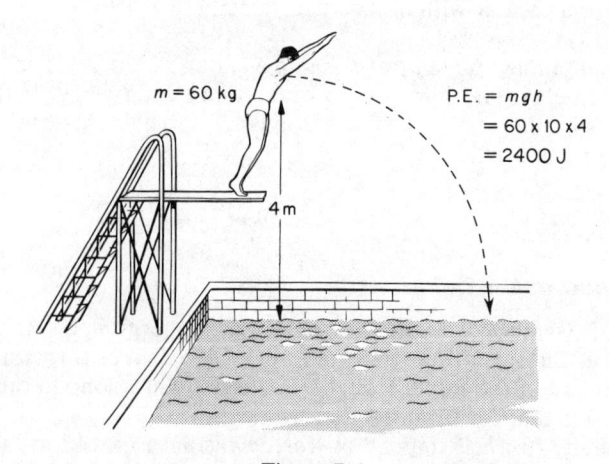

Figure 7.6.

(i) potential energy $= mgh = 1000 \times 10 \times 50 = 500\,000$ J
(ii) potential energy $mgh = 60 \times 10 \times 4 = 2400$ J

Kinetic energy

The *kinetic energy* of a moving object is the energy it has due to its motion. It can be shown (see the author's *Principles of Physics*, Hart-Davis) that the kinetic energy of a mass m moving with a velocity v is given by

$$\text{kinetic energy} = \tfrac{1}{2}mv^2$$

The kinetic energy is in J when m is in kg and v is in m/s.

Example

Calculate the kinetic energy of (i) a tennis ball of mass 0.1 kg moving with a velocity of 20 m/s, (ii) a car of mass 1000 kg moving with a velocity of 15 m/s.

(i) kinetic energy $= \tfrac{1}{2}mv^2 = \tfrac{1}{2} \times 0.1 \times 20^2 = 20$ J
(ii) kinetic energy $= \tfrac{1}{2}mv^2 = \tfrac{1}{2} \times 1000 \times 15^2 = 112\,500$ J

Transfer of energy

Energy can be transferred from potential energy to kinetic energy. For example, the stretched string of a bow in a bow-and-arrow has potential energy because the molecules of the string are further apart when the string is stretched. When the string is released, its potential energy is transferred to kinetic energy of the arrow.

Suppose the potential energy of the stretched string is 80 J and the mass m of the arrow is 0.05 kg. Then, at the instant the arrow is released, its kinetic energy, $\tfrac{1}{2}mv^2$, is 80 J, where v is the velocity of the arrow.

$$\therefore\ \tfrac{1}{2} \times 0.05 \times v^2 = 80$$

$$\therefore\ v^2 = \frac{80 \times 2}{0.05} = 3200$$

$$\therefore\ v = \sqrt{3200} = 57 \text{ m/s}$$

EXERCISE 24

Work; *energy* (where necessary, assume $g = 10$ m/s^2).

1. A car of mass 1000 kg is moving with a velocity of 20 m/s. Calculate its kinetic energy.

2. A load of weight 100 N is raised steadily through 5 m. Find (i) the work done, (ii) the gain in potential energy of the load.

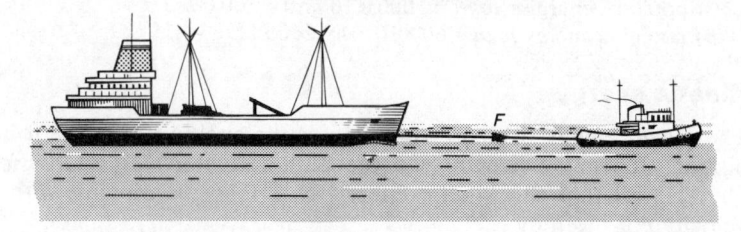

Figure 7.7.

3. A football of mass 0.4 kg is kicked so that it starts to move at 20 m/s. Calculate its initial kinetic energy.

4. A tug exerts a pull F of 2000 N and moves a ship 30 m in the direction of the pull (Figure 7.7). Calculate the work done.

5. A rocket of mass 20 kg reaches a height of 40 m and is then moving at a velocity of 100 m/s. Calculate its potential energy and kinetic energy at this height.

6. A model rocket on the ground is given an energy of 60 J, which lifts it vertically to a height h. Calculate h if the mass of the rocket is 0.5 kg.

7. In Question 6, what is the velocity of the rocket at the instant it starts to rise?

8. A ball of mass 0.2 kg is released from a height of 20 m. Using transfer of energy from potential to kinetic, calculate its velocity just before it hits the ground.

ELECTRICITY

Current; potential difference; resistance

If I is the current flowing in a resistance R, and V is the potential difference (p.d.) across R (Figure 7.8(a)), then

$$I = \frac{V}{R}, \quad V = IR, \quad R = \frac{V}{I}$$

Units I is in amperes (A), V is in volts (V), R is in ohms (Ω). So the current I flowing in a 10 Ω resistor when a p.d. of 12 V is connected across the resistor is (Figure 7.8(b))

$$I = \frac{V}{R} = \frac{12}{10} = 1.2 \text{ A}$$

APPLICATIONS OF FORMULAE IN PHYSICS

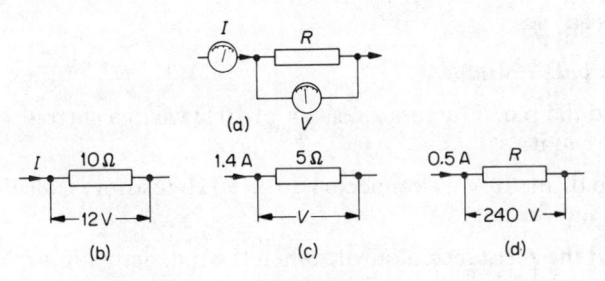

Figure 7.8.

If a current I of 1.4 A flows in a resistor of 5 Ω (Figure 7.8(c)), then the p.d. V across the resistor is given by

$$V = IR = 1.4 \times 5 = 7 \text{ V}$$

If the current I flowing in a resistor is 0.5 A when a p.d. V of 240 V is connected across it (Figure 7.8(d)), then the resistance R is given by

$$R = \frac{V}{I} = \frac{240}{0.5} = \frac{2400}{5} = 480 \text{ Ω}$$

Charge and electrolysis

If I is the current in a circuit, the quantity of *charge* Q which passes through a section in a time t is given by

$$Q = It$$

Units Q is in coulombs (C) when I is in amperes (A) and t is in seconds (s). So if $I = 3$ A and $t = 1$ min $= 60$ s, then

$$Q = 3 \times 60 = 180 \text{ C}$$

The *faraday* (F) is defined as the quantity of charge required to liberate 1 mole of any monovalent element in electrolysis, for example, 1 mole of silver (p. 141).

$$1 \text{ F} = 96\,500 \text{ C approximately}$$

Suppose a current of 2 A flows through a voltameter for 10 min or 600 s. Then the charge flowing is given by

$$Q = It = 2 \times 600 = 1200 \text{ C}$$
$$= \frac{1200}{96\,500} \text{ F}$$
$$= 0.012 \text{ F}$$

EXERCISE 25

Current, p.d., resistance

1. Find the p.d. V across a resistor of 10 Ω when a current I of 3 A flows in it.
2. A p.d. of 20 V is connected to a 50 Ω resistor. Calculate the current flowing.
3. Find the resistance of a wire when the p.d. across it is 2.8 V and the current flowing is 2 A.
4. Calculate the current flowing in a 4 Ω resistor when a p.d. of 2 V is connected to it.
5. Find the p.d. across a 3.6 Ω resistor when the current flowing in it is 0.5 A.

Charge

6. Find the charge Q in C passing a section of a wire in a time t of 5 min if the current I flowing is 1.5 A.
7. The charge flowing past a section of a conductor in 10 min is 1500 C. What current is flowing?
8. In electrolysis, 0.02 F passes through a voltameter when the current is 2 A. How long has the current flowed? (1 F = 96 500 C.)
9. 1 mole of silver has a mass of 108 g. What charge in F is required to liberate 2.7 g of silver in electrolysis?
10. 6.3 g of copper is liberated in electrolysis by 0.2 F. Calculate (to 2 s.f.) the mass per coulomb of copper liberated.

Series and parallel resistances; resistivity

Series and parallel resistances. If two resistances R_1 and R_2 are in series, their total resistance R is given by (Figure 7.9(a))

$$R = R_1 + R_2$$

So with $R_1 = 4 \, \Omega$ and $R_2 = 6 \, \Omega$,

$$R = 4 + 6 = 10 \, \Omega$$

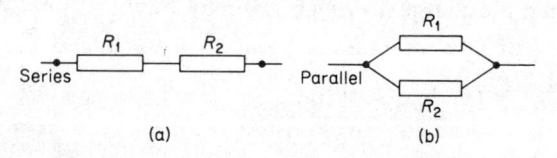

Figure 7.9.

APPLICATIONS OF FORMULAE IN PHYSICS

If R_1 and R_2 are in *parallel*, however (Figure 7.9(b)), then the total resistance R is now given by

$$\frac{1}{R} = \frac{1}{R_1} + \frac{1}{R_2}$$

So with $R_1 = 4\,\Omega$ and $R_2 = 12\,\Omega$,

$$\frac{1}{R} = \frac{1}{4} + \frac{1}{12} = \frac{3+1}{12} = \frac{4}{12}$$

Cross-multiplying,

$$4R = 12, \quad R = 3\,\Omega$$

Resistivity. The resistance R of a length l of a conductor of cross-sectional area A (Figure 7.10) is given by

$$R = \rho \frac{l}{A}$$

where ρ is the 'resistivity' of the conductor.

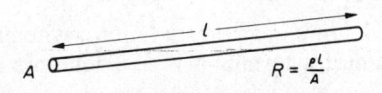

Figure 7.10.

Units R is in ohms (Ω) when ρ is in ohm metres (Ω m), l is in metres (m) and A is in metre^2 (m^2).

Suppose the diameter (gauge) of a wire is 0.60 mm. The radius $r = 0.30$ mm $= 0.3/1000$ m $= 0.3 \times 10^{-3}$ m, and so

area of cross-section, $A = \pi r^2 = \pi \times (0.3 \times 10^{-3})^2$
$$= \pi \times (3 \times 10^{-4})^2 = \pi \times 9 \times 10^{-8}$$
$$= 2.8 \times 10^{-7}\,\text{m}^2 \text{ (approx.)}$$

Suppose a Nichrome wire has a length of 4 m, a cross-sectional area of 2.8×10^{-7} m^2 and a resistivity $1 \times 10^{-6}\,\Omega$ m. Its resistance R is then given by

$$R = \rho\frac{l}{A} = 1 \times 10^{-6} \times \frac{4}{2.8 \times 10^{-7}}$$

$$= \frac{4}{2.8} \times 10^{-6-(-7)}$$

$$= \frac{4}{2.8} \times 10^1 = 14.3\,\Omega$$

EXERCISE 26

1. Wires of resistance $3\,\Omega$ and $4\,\Omega$ are joined (i) in series, (ii) in parallel. Find the total resistance in each case.

2. The combined resistance of two wires in parallel is $3\,\Omega$. If the resistance of one wire is $9\,\Omega$, calculate the resistance of the other wire.

3. A p.d. of 30 V is connected across two wires of $10\,\Omega$ and $15\,\Omega$ arranged (i) in series, and then (ii) in parallel. Find the current flowing from the battery in each case.

4. Calculate the resistance R of a length l of 10 m of wire of cross-sectional area A of $4 \times 10^{-8}\,m^2$ and resistivity ρ of $5 \times 10^{-7}\,\Omega\,m$.

5. The filament of a lamp has a length of 5 m, a cross-sectional area of $1 \times 10^{-7}\,m^2$ and a resistivity of $2 \times 10^{-6}\,\Omega\,m$. Find its resistance.

6. A copper cable has a length of 50 m, a diameter of 2 mm and a resistivity of $2 \times 10^{-7}\,\Omega\,m$. Calculate the resistance of the cable.

Electrical energy

Suppose a current I flows in an electric motor when it is working, and the p.d. across the motor terminals is V. Then, in a time t,

electric energy used, $\quad W = IVt$

The energy W is in joules (J) when I is in amperes (A), V is in volts (V) and t is in seconds (s), as in Figure 7.11(a).

If the current in the motor is 3 A and the p.d. across the motor is 240 V when it is used to drive a lathe, then, in a time of 1 min or 60 s,

energy used by motor, $\quad IVt = 3 \times 240 \times 60$

$$= 43\,200\,J$$

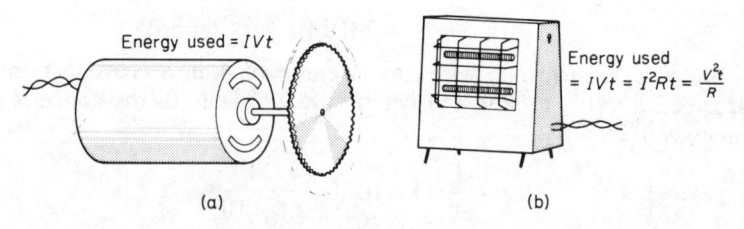

Figure 7.11.

Heating effect of current

When a current I flows in an electric fire element of resistance R, *all* the energy supplied is transferred to heat energy (Figure 7.11(b)). In

APPLICATIONS OF FORMULAE IN PHYSICS

this special case, the p.d. $V = IR$ (p. 68). So

$$\text{heat produced,} \quad IVt = I \times IR \times t = I^2Rt$$

Also, since $I = V/R$, we can write

$$\text{heat produced,} \quad IVt = \frac{V}{R} \times V \times t = \frac{V^2 t}{R}$$

In these formulae the heat produced is in joules (J) when R is in ohms (Ω), I in amperes (A) and V in volts (V). So if the resistance is 120 Ω and a current of 2 A flows in it for 1 min or 60 s,

$$\text{heat produced} = I^2Rt = 2^2 \times 120 \times 60$$
$$= 4 \times 120 \times 60 = 28\,800 \text{ J}$$

Electrical power

The *power* of any machine is the energy used per second. Now, from previous, the energy used by a machine is IVt. So

$$\text{power,} \quad P = \frac{\text{energy}}{\text{time}}$$

$$= \frac{IVt}{t} = IV$$

The power, P, is in *watts* (W) when I is in amperes (A) and V is in volts (V). In a resistor R we have $V = IR$. So in this special case,

$$P = IV = I^2 R$$

Also

$$P = IV = \frac{V}{R} \times V = \frac{V^2}{R}$$

Example

Calculate the power of an electric lamp which has a filament resistance of 480 Ω when working, and a current of $\frac{1}{2}$ A flowing through it.

$$P = I^2 R = (\tfrac{1}{2})^2 \times 480$$
$$= \tfrac{1}{4} \times 480 = 120 \text{ W}$$

EXERCISE 27

1. An electric carpet sweeper has a current I of 1.1 A when a p.d. V of 240 V is connected across it (Figure 7.12(a)). What energy is used in a time of 2 min?

74 BASIC MATHEMATICS FOR SCIENCE

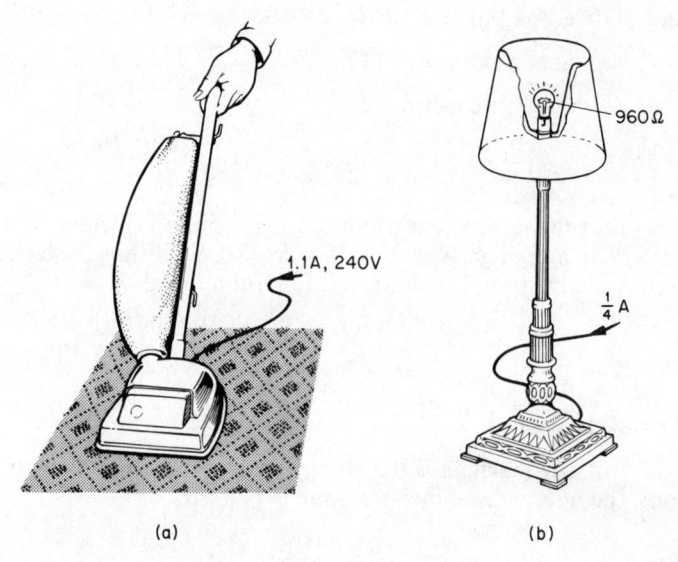

(a) (b)

Figure 7.12.

2. An electric heater has a 100 Ω resistance and uses a current of $\frac{1}{2}$ A. What energy is used in 30 s?

3. An electric motor has a p.d. of 12 V and carries a current of 2 A. What energy is used in 1 min?

4. A lamp filament has a resistance of 960 Ω when used and a current of $\frac{1}{4}$ A (Figure 7.12(b)). What is the power of the lamp?

5. A radio resistor carries a current of 0.1 A when the p.d. across it is 10 V. What is its resistance and the power used?

6. An electric heater has a resistance of 100 Ω when used on a p.d. of 50 V. What is its power?

HEAT

Heat capacity

The *heat capacity*, C, of an object such as a block of metal or a tank is the heat required to raise its temperature by 1 K (1°C).

Suppose the temperature of a metal block rises from 10°C to 15°C, a rise θ of 5 K (5°C), when a quantity of heat Q of 500 J is transferred to it. Then

$$\text{heat capacity,} \quad C = \frac{\text{heat gained, } Q}{\text{temperature rise, } \theta}$$

$$= \frac{500 \text{ J}}{5 \text{ K}} = 100 \text{ J/K}$$

APPLICATIONS OF FORMULAE IN PHYSICS 75

If the temperature of the block falls by 3 K, the heat lost by it is $100 \times 3 = 300$ J. Generally, the heat gained or lost from an object of heat capacity C when its temperature changes by θ is given by

$$Q = C\theta$$

Specific heat capacity

The *specific heat capacity*, c, of a substance is the heat required to raise the temperature of 1 kg of it by 1 K (1°C). In the case of a metal block,

$$c = \frac{\text{heat capacity of block, } C}{\text{mass of block in kg, } m}$$

So if the metal block on p. 74 had a mass of 0.2 kg, the specific heat capacity of the metal is given by

$$c = \frac{100 \text{ J/K}}{0.2 \text{ kg}} = 500 \text{ J/kg K}$$

This means that 1 kg of the metal rises by 1 K (1°C) when a quantity of heat Q of 500 J is transferred to or gained by the metal. Generally, if a mass m kg rises (or falls) in temperature by θ K, the heat gained (or lost) is given in joules by

$$Q = mc\theta$$

Water has a high specific heat capacity of about 4200 J/kg K. Soil has a much lower specific heat capacity. For this reason, the soil becomes warmer than the water in a neighbouring pond when exposed to sunlight.

Example

Calculate (i) the heat gained by 0.2 kg of aluminium, specific heat capacity 900 J/kg K, when its temperature rises from 15°C to 19°C, (ii) the total heat lost by 0.5 kg of water inside an aluminium calorimeter (can) of mass 0.1 kg when their temperature falls from 12°C to 10°C.

(i)   heat gain, $Q = mc\theta$
$$= 0.2 \times 900 \times (19 - 15) = 720 \text{ J}$$

(ii)   heat lost by water, $Q = mc\theta$
$$= 0.5 \times 4200 \times (12 - 10) = 4200 \text{ J}$$

               heat lost by can, $Q = mc\theta$
$$= 0.1 \times 900 \times (12 - 10) = 180 \text{ J}$$

$\therefore$ total heat lost $= 4200 + 180 = 4380$ J

Electrical heating

A small electric heater may produce 50 W, which means 50 J/s. If it is used for 4 min, which is 4×60 or 240 s,

heat produced, $Q = 50 \times 240 = 12\,000$ J

If this heater had been used for warming a mass m of 0.4 kg of water with specific heat capacity c of 4200 J/kg K, the temperature rise θ would be given by

$$Q = mc\theta$$

or

$$\theta = \frac{Q}{mc}$$

$$= \frac{12\,000}{0.4 \times 4200} = 7 \text{ K } (7°C)$$

Specific latent heat of vaporization (evaporation)

The *specific latent heat of vaporization* of a liquid is the heat required to change 1 kg of it from liquid to vapour *at the boiling point*. Water requires about 2 200 000 J (2.2×10^6 J) to change 1 kg of it to vapour (steam). So

specific latent heat, $l = 2\,200\,000$ J/kg

Generally, the heat Q required to change m kg of a liquid to vapour *at the boiling point* is given by

$$Q = ml$$

So 0.1 kg of water, brought to its boiling point of 100°C, will all evaporate if it is given a quantity of heat Q where

$$Q = ml = 0.1 \times 2\,200\,000$$
$$= 220\,000 \text{ J}$$

This is also the *heat given up* when the 0.1 kg of steam (vapour) condenses to water *at 100°C*. If the water formed cools to say 30°C, an additional amount of heat is given up, which is

$$Q = mc\theta = 0.1 \times 4200 \times (100 - 30)$$
$$= 29\,400 \text{ J}$$

So if 0.1 kg of steam condenses to water at 30°C, the total heat given up is

$$Q = ml + mc\theta$$
$$= 220\,000 + 29\,400 = 249\,400 \text{ J}$$

APPLICATIONS OF FORMULAE IN PHYSICS 77

The latent heat given up when steam condenses to water at 100°C is much greater than the heat given up when the water formed cools to 30°C. This is why steam can produce more severe scalds than hot water.

Specific latent heat of fusion

The specific latent heat of fusion, l, of a solid is the heat required to change 1 kg of it from solid to liquid *at the melting point*.

Ice has a specific latent heat of fusion of about 340 000 J/kg. So the heat required to melt 0.5 kg of ice to water *at $0°C$* is

$$Q = ml = 0.5 \times 340\,000 = 170\,000 \text{ J}$$

If the water formed rises in temperature from 0°C to 8°C, or by 8 K, then an additional amount of heat Q' is required which is

$$Q' = mc\theta = 0.5 \times 4200 \times 8$$
$$= 16\,800 \text{ J}$$

So the total heat needed to change 0.5 kg of ice to water at 8°C is

$$Q + Q' = 170\,000 + 16\,800 = 186\,800 \text{ J}$$

EXERCISE 28

(Where necessary, assume the specific heat capacity of water = 4200 J/kg K.)

1. An iron tank has a mass 2 kg. Find the heat needed to raise its temperature from 10°C to 15°C (c for iron = 440 J/kg K).
   What is the heat capacity of the tank?

2. An aluminium kettle of mass 0.1 kg contains 0.3 kg of water at 15°C. Calculate the heat required to raise the water to 100°C. (c for aluminium = 900 J/kg K.)

3. In an experiment, a 50 W heating coil, used for 4 min, warms a metal block of mass 2 kg from 10°C to 18°C. Find (i) the heat supplied, (ii) the heat capacity of the metal block, (iii) the specific heat capacity of the metal.

4. Calculate the mass of water which rises in temperature from 12°C to 17°C when 84 000 J of heat is transferred to it.
   What is the total heat required for the same temperature rise if the water is inside a copper vessel of mass 0.5 kg? (c for copper = 400 J/kg K.)

Latent Heat (Where necessary, assume the specific latent heat of vaporization of water is 2 200 000 J/kg (2.2×10^6 J/kg) and the specific latent heat of fusion of ice is 330 000 J/kg (3.3×10^5 J/kg).)

5. A burner supplies 440 J/s. How long will it take to evaporate 0.1 kg of water at 100°C to steam at the same temperature?

6. Calculate the quantity of heat given up when hot iron, mass 0.2 kg and temperature 150°C, is placed in the middle of a large block of ice at 0°C (c for iron = 440 J/kg K).
What mass of ice is melted by the iron?

7. 0.01 kg of steam at 100°C condenses to water at 60°C after it is directed by a jet on to a metal block of mass 1 kg at 10°C.
Calculate (i) the heat given up when the steam first condenses to water at 100°C, (ii) the total heat given up by the steam in condensing to water at 60°C, (iii) the specific heat capacity of the metal, if all the heat in (ii) is given to it to raise its temperature to 60°C.

8. Calculate the heat transferred to it when 0.01 kg of ice melts (i) to water at 0°C, (ii) to water at 5°C.
What is the drop in temperature of 0.2 kg of a liquid of specific heat capacity 2000 J/kg K if this ice is dropped into it and forms water at 5°C?

GEOMETRICAL OPTICS

Curved mirrors and lenses

When an object is placed in front of a curved spherical mirror or a thin lens, the image position can be calculated from the formula

$$\frac{1}{v} + \frac{1}{u} = \frac{1}{f}$$

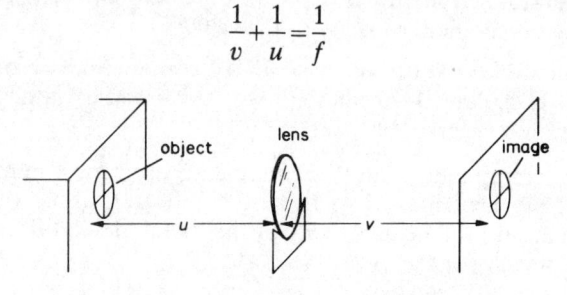

Figure 7.13.

Here v is the image distance, u is the object distance and f is the focal length of the mirror or lens (Figure 7.13). This formula applies only if the 'real-is-positive' sign convention is used, that is, a real image or object distance is given a + sign in front of its numerical value, and a virtual image or object distance is given a − sign, and f is +ve if the principal focus is real or −ve if it is virtual.

Curved mirrors and f

Here the image is produced by *reflection*. Parallel rays incident on a *concave* mirror are reflected to a focus F in front of the mirror (Figure 7.14(a)), so that F is real. Hence the focal length f, the distance from F to the mirror, is +ve.

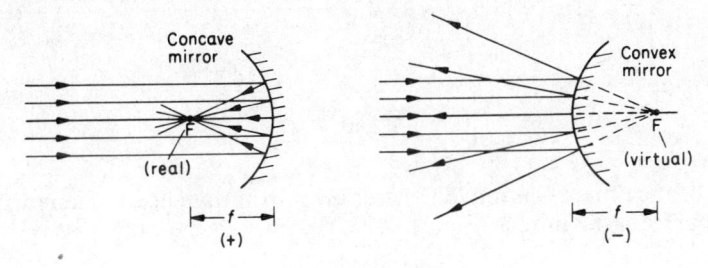

Figure 7.14.

Parallel rays incident on a *convex* mirror, however, are reflected as if they diverged from a point F *behind* the mirror (Figure 7.14(b)). So F is virtual. Hence the focal length f of a convex mirror is −ve.

Object and image calculations

Concave mirror **1.** Suppose an object O is placed 20 cm from a concave mirror of focal length 12 cm (Figure 7.15(a)). Then

$$u = +20 \text{ cm (real object)} \quad \text{and} \quad f = +12 \text{ cm}$$

Substituting in

$$\frac{1}{v} + \frac{1}{u} = \frac{1}{f}$$

we obtain

$$\frac{1}{v} + \frac{1}{(+20)} = \frac{1}{(+12)}$$

So

$$\frac{1}{v} = \frac{1}{12} - \frac{1}{20} = \frac{5-3}{60}$$

$$= \frac{2}{60}$$

Cross-multiplying,

$$2v = 60, \quad v = \frac{60}{2} = 30 \text{ cm}$$

Since $v = +30$ cm, the image is real and so it is formed 30 cm in front of the mirror, as shown.

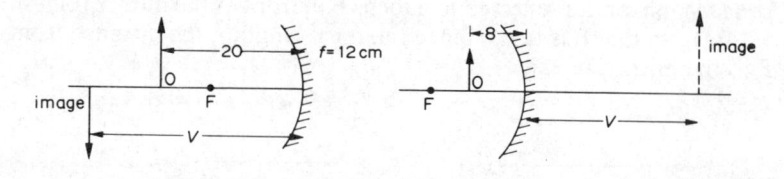

Figure 7.15.

2. If the object O is moved nearer the mirror than its focal length, say 8 cm from the mirror (Figure 7.15(b)). Then, as before,

$$u = +8 \text{ cm} \quad \text{and} \quad f = +12 \text{ cm}$$

From

$$\frac{1}{v} + \frac{1}{u} = \frac{1}{f}$$

$$\frac{1}{v} + \frac{1}{(+8)} = \frac{1}{(+12)}$$

So

$$\frac{1}{v} = \frac{1}{12} - \frac{1}{8} = \frac{2-3}{24} = -\frac{1}{24}$$

$$v = -24 \text{ cm}$$

This time the image distance is negative, so the image is *virtual*. This means that the image is formed 24 cm *behind* the mirror, as shown.

Convex mirror With a convex mirror, we must remember that f is *negative*. For example, suppose an object O is placed 20 cm from a convex mirror of focal length 30 cm (Figure 7.16). Then $u = +20$ cm, and $f = -30$. Substituting in

$$\frac{1}{v} + \frac{1}{u} = \frac{1}{f}$$

Figure 7.16.

APPLICATIONS OF FORMULAE IN PHYSICS

we obtain

$$\frac{1}{v} + \frac{1}{(+20)} = \frac{1}{(-30)}$$

So

$$\frac{1}{v} + \frac{1}{20} = -\frac{1}{30}$$

$$\frac{1}{v} = -\frac{1}{30} - \frac{1}{20} = \frac{-2-3}{60} = -\frac{5}{60}$$

$$\frac{1}{v} = -\frac{1}{12}, \quad v = -12$$

Since v is negative, the image is 12 cm *behind* the mirror, as shown.

Lenses and *f*

Parallel rays incident on a *converging lens* are *refracted* to a focus F beyond the lens (Figure 7.17(d)). So F is real and the focal length f, the distance from the lens to F, is a +ve distance.

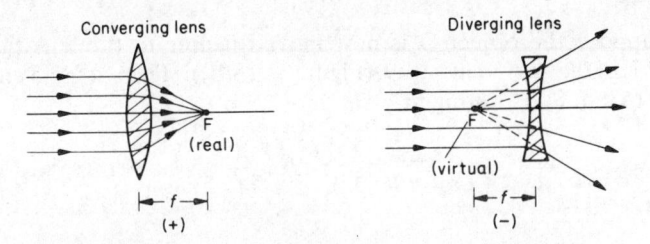

Figure 7.17.

Parallel rays incident on a *diverging lens* are refracted so that they appear to diverge from a focus F behind the lens (Figure 7.17(b)). So F is virtual. Hence the focal length f of a diverging lens is −ve.

Object and image calculations with lenses

Converging lens **1.** Suppose an object O is placed 30 cm from a converging lens of focal length 15 cm (Figure 7.18(a)). Then $u = +30$ cm and f is +15 cm. Substituting in

$$\frac{1}{v} + \frac{1}{u} = \frac{1}{f}$$

gives

$$\frac{1}{v} + \frac{1}{(+30)} = \frac{1}{(+15)}$$

$$\therefore \frac{1}{v} + \frac{1}{30} = \frac{1}{15}$$

$$\frac{1}{v} = \frac{1}{15} - \frac{1}{30} = \frac{2-1}{30} = \frac{1}{30}$$

$$\therefore v = 30$$

Since v is +ve, the image is real. So the image is formed beyond the lens and 30 cm from it, as shown.

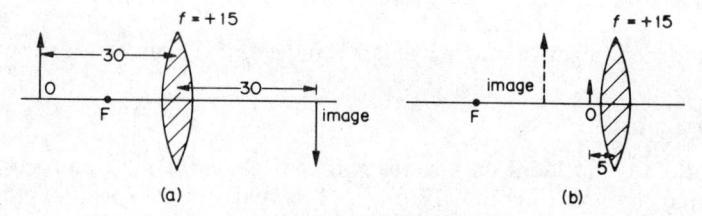

Figure 7.18.

2. Suppose the object O is now moved nearer to the lens than its focal length, say 5 cm away (Figure 7.18(b)). Then $u = +5$ cm and $f = +15$ cm. Substituting in

$$\frac{1}{v} + \frac{1}{u} = \frac{1}{f}$$

gives

$$\frac{1}{v} + \frac{1}{(+5)} = \frac{1}{(+15)}$$

$$\therefore \frac{1}{v} + \frac{1}{5} = \frac{1}{15}$$

$$\therefore \frac{1}{v} = \frac{1}{15} - \frac{1}{5} = \frac{1-3}{15} = -\frac{2}{15}$$

$$2v = -15, \qquad v = -\frac{15}{2} = -7.5 \text{ cm}$$

Since v is −ve, the image is virtual. It is therefore formed 7.5 cm from the lens on the same side as the object, as shown.

APPLICATIONS OF FORMULAE IN PHYSICS 83

Diverging lens In this case f is $-$ve. Suppose an object O is placed 12 cm from a diverging lens of focal length 8 cm (Figure 7.19). Then $u = +12$ cm and $f = -8$ cm. Substituting in the lens equation $1/v + 1/u = 1/f$ gives

$$\frac{1}{v} + \frac{1}{(+12)} = \frac{1}{(-8)}$$

$$\therefore \quad \frac{1}{v} + \frac{1}{12} = -\frac{1}{8}$$

$$\therefore \quad \frac{1}{v} = -\frac{1}{8} - \frac{1}{12} = \frac{-3-2}{24} = -\frac{5}{24}$$

So

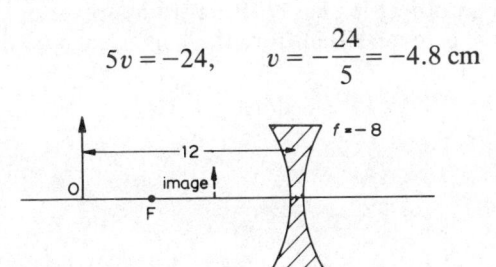

$$5v = -24, \quad v = -\frac{24}{5} = -4.8 \text{ cm}$$

Figure 7.19.

So the image is virtual and therefore formed 4.8 cm from the lens on the same side as the object, as shown.

EXERCISE 29

Curved mirrors

1. An object is placed 25 cm (u) from a concave mirror of focal length 15 cm (f). Find the image distance (v).

2. In Question 1, find the new image distance if the object is moved to a position 10 cm from the mirror. How does the image differ from that in Question 1?

3. When an object is placed 12 cm from a concave mirror, a real image is obtained 24 cm from the mirror. Calculate the focal length.

4. An object is placed 48 cm from a *convex* mirror of focal length 12 cm. Find the image distance. Is the image real or virtual?

5. When an object is placed 40 cm from a convex mirror, a virtual image is formed 8 cm from the mirror. Find the focal length of the mirror.

Lenses

6. A converging lens has a focal length of 10 cm. Find the image distance when an object is placed 30 cm from the lens.

7. An object is placed 12 cm from a converging lens and a real image is formed 24 cm from the lens. Calculate the focal length of the lens.

8. With the lens of Question 7, where must an object be placed to form a *virtual* image 24 cm from the lens?

9. An object is placed 16 cm from a *diverging* lens of focal length 12 cm. Find the image distance.

10. When an object is placed 20 cm from a diverging lens, a *virtual* image is formed 10 cm from the lens. Calculate the focal length of the lens.

8 Changing Formulae: Mechanics; Electricity; Heat

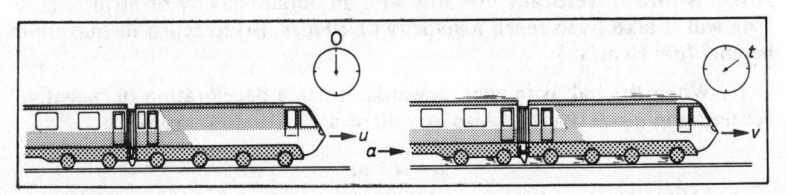

Figure 8.1.

MECHANICS

Linear equation

We often need to change a formula into a more useful form. For example, we may want t from the equation of motion $v = u + at$ (p. 62).

In this case we can imagine t to be an 'unknown' quantity, and the other letters v, u and a to be 'known' quantities. We then have to solve a linear equation for t. As previously shown, we take the unknown quantity t to one side of the equation and the known quantities v, u and a to the other side. So

$$v = u + at$$
$$\therefore v - u = at$$

or
$$at = v - u$$
$$\therefore t = \frac{v - u}{a}$$

If you have any difficulty with solving an equation with 'known' quantities in the form of letters, as here, simply replace the letters by numbers and re-write the equation. For example, in place of $v = u + at$, write $7 = 3 + 2t$. Then

$$7 - 3 = 2t$$

85

so

$$t = \frac{7-3}{2}$$

This confirms that

$$t = \frac{v-u}{a}$$

Example

A ball is thrown vertically upwards with an initial velocity of 30 m/s. How long will it take (i) to reach a velocity of 10 m/s, (ii) to reach its maximum height? ($g = 10$ m/s^2.)

(i) When the ball is moving upwards, it has a deceleration or 'negative' acceleration $a = -10$ m/s^2. Also, $u = 30$ m/s, $v = 10$ m/s. From

$$v = u + at$$
$$\therefore \quad 10 = 30 - 10t$$
$$\therefore \quad 10t = 30 - 10 = 20$$
$$t = \frac{20}{10} = 2 \text{ s}$$

(ii) At its maximum height, the velocity $v = 0$. From

$$v = u + at$$
$$\therefore \quad 0 = 30 - 10t$$
$$\therefore \quad 10t = 30$$
$$t = \frac{30}{10} = 3 \text{ s}$$

Equation with squares

Suppose u is needed from the equation of motion $v^2 = u^2 + 2as$ (p. 62). In this case u is the 'unknown' and v, a and s are the 'known' quantities. So

$$v^2 - 2as = u^2$$

or

$$u^2 = v^2 - 2as$$

Take the *square root* of both sides of the equation. Then

$$u = \sqrt{v^2 - 2as}$$

Equation with square root

Suppose g is needed from the equation for a simple pendulum (Figure 8.2), which is

$$T = 2\pi\sqrt{\frac{l}{g}}$$

Square both sides of the equation to get rid of the square root. Then

$$T^2 = \frac{4\pi^2 l}{g}$$

Cross-multiply,

$$\therefore \quad gT^2 = 4\pi^2 l$$

$$\therefore \quad g = \frac{4\pi^2 l}{T^2}$$

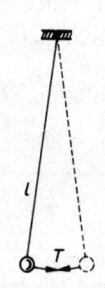

Figure 8.2.

Exploding objects

When a stationary object explodes into two parts, each part flies off in opposite directions with *equal momentum*. This is because the force on one part is equal to the force on the other part, from Newton's Law of Action and Reaction, and each force acts for the same time (see p. 64).

Suppose one part A has a mass of 4 kg and flies off with a velocity of 60 m/s, and the other part B has a mass of 5 kg and velocity v. Then

$$\text{momentum of } B = \text{momentum of } A$$

$$\therefore \quad 5 \times v = 4 \times 60$$

$$v = \frac{4 \times 60}{5} = 48 \text{ m/s}$$

Colliding objects

For a similar reason, when two objects X and Y collide

total momentum of X and Y before collision = total momentum after collision

In this case the force on X due to the impact is equal and opposite to the force on Y, from Newton's Law, and each force acts for the same time. So if the momentum change of X is $+4$ units, for example, the momentum change of Y is -4 units. Added together, the total change of $+4-4=0$. So the total momentum of X and Y is unchanged by the collision. This is called the *Principle of the Conservation of Linear Momentum*. No forces other than those due to impact are concerned, otherwise the total momentum is not constant.

Examples

1. A ball A, mass 2 kg and moving with velocity 10 m/s, collides head-on with a stationary ball B of mass 3 kg. After impact, A and B stick together and move with the same velocity v. Calculate v.

Momentum of A and B before collision $= 2 \times 10 = 20$ m/s, since the momentum of $B = 0$.

Momentum of A and B after collision $= (2+3)v = 5v$

$$\therefore \quad 5v = 20$$

$$v = \frac{20}{5} = 4 \text{ m/s}$$

2. A ball X of mass 4 kg and moving with velocity 8 m/s collides head-on with a ball Y of mass 3 kg moving in the *opposite* direction with a velocity of 2 m/s. After collision, X moves in the same direction with a reduced velocity v, and Y now moves in the *same* direction as X with a velocity of 2 m/s. Calculate v.

The total momentum of X and Y in the direction of X initially

$$= (4 \times 8) - (3 \times 2) = 32 - 6$$

since the momentum of Y is in the *opposite* direction to that of X.

After collision, since the momentum of Y is in the *same* direction as that of X,

total momentum in direction of $X = (4 \times v) + (3 \times 2) = 4v + 6$

From the conservation of momentum,

$$\therefore \quad 4v + 6 = 32 - 6$$

$$4v = 32 - 6 - 6 = 20$$

$$v = \frac{20}{4} = 5 \text{ m/s}$$

CHANGING FORMULAE: MECHANICS; ELECTRICITY; HEAT   89

EXERCISE 30
(Where necessary, assume $g = 10$ m/s^2.)

1. Find a from $v = u + at$.
2. Find l from $T = 2\pi\sqrt{\dfrac{l}{g}}$.
3. If $A = \pi r^2$, find r.
4. Find p from $q^2 = p^2 - r^2$.
5. From $F = ma$, find a.
6. In $p = h\rho g$, find h.
7. Find v from $mu = Mv + p$.
8. In $s = \frac{1}{2}gt^2$, find t.
9. If $V = \frac{4}{3}\pi r^3$, find r.
10. From $s = ut + \frac{1}{2}at^2$, find a.
11. A tennis ball is hit vertically upwards with an initial velocity u of 20 m/s. After a time t of 0.5 s, its velocity v is 15 m/s. Calculate the deceleration or 'negative acceleration' a of the ball from $v = u + at$.
12. Calculate the length l of a simple pendulum of period $T = 1.0$ s assuming $g = 10$ m/s^2, if $T = 2\pi\sqrt{l/g}$.
13. The velocity u of a train is initially 15 m/s and it accelerates steadily to a velocity v of 30 m/s over a distance s of 225 m. Find the acceleration a if $v^2 = u^2 + 2as$.
14. A cricket ball of mass m of 0.15 kg is hit into the air. At one stage it has a kinetic energy E of 30 J. Calculate its velocity at this stage if $E = \frac{1}{2}mv^2$.
15. In Question 14, the cricket ball has also potential energy due to its height above the ground. If the potential energy E' is 12 J, find the height h from $E' = mgh$, where g is 10 m/s^2.
16. A man of mass 60 kg jumps on to the shore from a boat of mass 150 kg with a velocity of 6 m/s. With what velocity does the boat start to move backwards? (*Use momentum.*)
17. A skater X of mass 60 kg, moving with a velocity of 10 m/s, collides head-on with another skater Y of mass 40 kg.
    (i) If Y is stationary, and X and Y both move together after impact, calculate their common velocity. (ii) If Y is moving in the *opposite* direction to X with a velocity of 5 m/s at the instant of collision, and X continues to move in the same direction with a velocity of 2 m/s after collision, find the velocity of Y. (*Use momentum.*)
18. A falling object of mass m of 5 kg meets air resistance. This reduces its acceleration a to 8 m/s^2 from a value g of 10 m/s^2. Find the resistance F from $mg - F = ma$, where F is in newtons, N.

19. The pressure p of water at a depth h is $1.5 \times 10^5 \, \text{N/m}^2$. If $p = h\rho g$, where ρ is the density of water, $1000 \, \text{kg/m}^3$, and g is $10 \, \text{m/s}^2$, calculate h.

20. A water pump, used to form a fountain spray, raises a mass m of 50 kg of water every minute through a height h. If the average power P needed to raise the water is 50 W, find h from $Pt = mgh$, where t is the time in seconds and g is $10 \, \text{m/s}^2$.

ELECTRICITY

Current; p.d.; resistance

Resistance R is defined from the relation (Figure 8.3)

$$R = \frac{V}{I} \tag{1}$$

Cross-multiply to find I. Then

$$IR = V$$

$$\therefore \quad I = \frac{V}{R} \tag{2}$$

To find V from (2), cross-multiply. So

$$V = IR \tag{3}$$

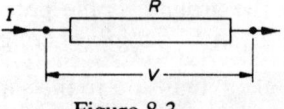

Figure 8.3.

Equations (1), (2), and (3) are formulae for R, I and V respectively.

Power

Power, $P = IV$ (Figure 8.4)

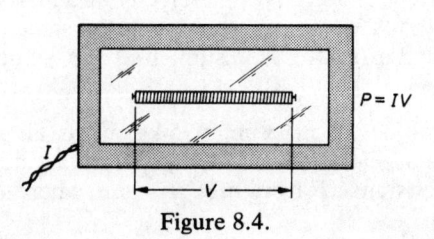

Figure 8.4.

CHANGING FORMULAE: MECHANICS; ELECTRICITY; HEAT

So
$$I = \frac{P}{V}$$

In the case of a resistance R,
$$P = I^2 R$$

So
$$I^2 = \frac{P}{R}$$
$$\therefore \quad I = \sqrt{\frac{P}{R}}$$

Also
$$P = \frac{V^2}{R}$$

To find R, cross-multiply. Then
$$PR = V^2$$
$$\therefore \quad R = \frac{V^2}{P}$$

Example

The power P of a lamp is 100 W when it is used on a 240 V mains. Calculate the resistance R of the lamp filament and the current I flowing when it is used.

We have
$$P = \frac{V^2}{R}$$
$$\therefore \quad 100 = \frac{240^2}{R}$$
$$\therefore \quad 100\,R = 240^2$$
$$\therefore \quad R = \frac{240^2}{100} = \frac{57\,600}{100} = 576\,\Omega$$

Also,
$$P = IV$$
$$\therefore \quad 100 = I \times 240$$
$$\therefore \quad I = \frac{100}{240} = 0.4 \text{ A (approx.)}$$

Resistivity

The resistance R of a wire of length l, cross-sectional area A and resistivity ρ is given by (Figure 8.5)

$$R = \frac{\rho l}{A}$$

To find l, we have

$$RA = \rho l$$

So

$$l = \frac{RA}{\rho} \tag{1}$$

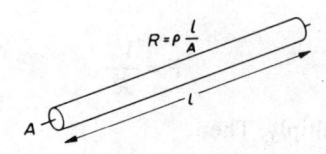

Figure 8.5.

The area of cross-section A is πr^2, where r is the radius of the wire. So

$$R = \frac{\rho l}{\pi r^2}$$

If r is needed, we first cross-multiply:

$$\pi r^2 R = \rho l$$

So

$$r^2 = \frac{\rho l}{\pi R}$$

$$\therefore \quad r = \sqrt{\frac{\rho l}{\pi R}} \tag{2}$$

Example

Find the length l of Manganin (resistance) wire which has a resistance of 20 Ω, if the diameter of the wire is 0.4 mm. (Resistivity of Manganin $\rho = 1 \times 10^{-6}$ Ω m.)

We have

$$R = 20 \, \Omega, \qquad \rho = 1 \times 10^{-6} \, \Omega \, \text{m}$$

CHANGING FORMULAE: MECHANICS; ELECTRICITY; HEAT

and
$$A = \pi r^2 = \pi \times (0.2 \times 10^{-3})^2 = \pi \times 4 \times 10^{-8} \text{ m}^2$$
From
$$R = \frac{\rho l}{A}$$
$$l = \frac{R \times A}{\rho}$$
$$= \frac{20 \times 3.14 \times 4 \times 10^{-8}}{1 \times 10^{-6}}$$
$$= 2.5 \text{ m}$$

E.m.f. and internal resistance

When a battery of e.m.f. (electromotive force) E and internal resistance r is connected to an external resistance R (Figure 8.6), a current I flows which is given by

$$I = \frac{E}{R+r}$$

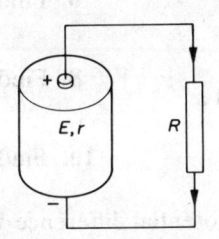

Figure 8.6.

To find E, cross-multiply. So

$$E = I(R+r)$$

Suppose we need r. Then we remove the brackets and solve the equation for r. So

$$E = IR + Ir$$
$$\therefore \quad Ir = E - IR$$
$$\therefore \quad r = \frac{E - IR}{I}$$

94   BASIC MATHEMATICS FOR SCIENCE

Example

A battery of e.m.f. 6 V supplies a current of 0.5 A when connected to a resistance of 10 Ω. Calculate the internal resistance of the battery.

If r is the internal resistance, then $I = \dfrac{E}{R+r}$

$$\therefore \ 0.5 = \dfrac{6}{10+r}$$

$$\therefore \ 0.5(10+r) = 6$$

$$\therefore \ 5 + 0.5r = 6$$

$$0.5r = 6 - 5 = 1$$

$$r = \dfrac{1}{0.5} = 2 \ \Omega$$

EXERCISE 31

1. If $R = \dfrac{\rho l}{A}$, find ρ.

2. From $I = \dfrac{V}{R}$, find R.

3. If $I = \dfrac{E}{R+r}$, find R.

4. Find I from $P = I^2 R$.

5. If $R = \dfrac{\rho l}{A}$, find l.

6. Find V from $P = \dfrac{V^2}{R}$.

7. If $W = IVt + w$, find t.

8. Find r from $I = \dfrac{2E}{R+2r}$.

9. If $R = \dfrac{\rho l}{\pi r^2}$, find l.

10. Find r from $\dfrac{E}{R+r} = \dfrac{V}{R}$.

11. In Figure 8.7(a), a potential difference V of 4.5 V is connected to a resistance R and a current I of 1.5 A flows. Calculate R.

12. The same potential difference of 4.5 V as in Question 11 is now connected across a 9 Ω resistance (Figure 8.7(b)). What current I flows?

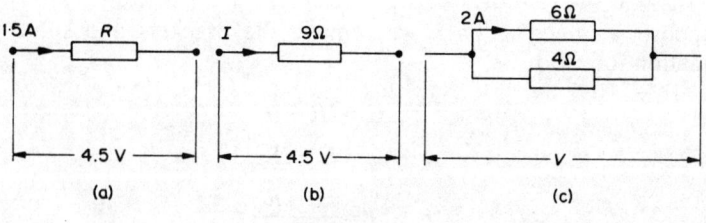

Figure 8.7.

CHANGING FORMULAE: MECHANICS; ELECTRICITY; HEAT   95

13. In Figure 8.7(c), calculate the potential difference V across the 6 Ω resistor if a current I of 2 A flows in it.
    Use the same value of V to find the current I flowing in the lower resistor R of 4 Ω.

14. When a battery with an e.m.f. E of 4 V and internal resistance r is joined to an 8 Ω resistance R, a current I of 0.4 A flows. Calculate r if $I = E/(R+r)$.

15. Using the formula for I in Question 14, find the e.m.f. E of a battery of internal resistance r of 3 Ω if a current I of 0.5 A flows when a resistance R of 15 Ω is connected to the battery.

16. An electric kettle uses a power P of 1000 W when the voltage V connected to it is 250 V. Calculate the resistance R of the heating element if $P = V^2/R$.

17. In Question 16, using $P = I^2R$, find the current I flowing in the heating element. Check your result from $I = V/R$.

18. A heating element has a resistance R of 960 Ω when it uses a power P of 60 W. Calculate the voltage V of the supply if $P = V^2/R$.

19. A wire has a resistance R of 2 Ω, a constant cross-sectional area A of 6×10^{-7} m^2, and a resistivity ρ of 10^{-6} Ω m. Calculate the length l of the wire from $R = \rho l/A$.

20. Find the resistivity ρ of a metal if a length l of 5 m and cross-sectional area A of 5×10^{-7} m^2 has a resistance R of 20 Ω, using the formula in Question 19.

HEAT

Heat capacity; specific heat capacity

From p. 75, to which the reader should refer, the heat Q transferred to an object of heat capacity C when its temperature rises by θ is given by

$$Q = C\theta \qquad (1)$$

Q is in joules (J), C is in J/K and θ is in K (or °C).
If the mass of the object is m and its specific heat capacity is c, then

$$Q = mc\theta \qquad (2)$$

Q is in J, m is in kg, c is in J/kg K, θ is in K (or °C).
From (2), the temperature rise θ is given by

$$\theta = \frac{Q}{mc}$$

Example

When oil is poured into a tank, a quantity of heat Q of 20 000 J is transferred from the oil to the tank. The oil temperature then drops from 16°C to 14°C. Calculate the mass of the oil if its specific heat capacity c is 2500 J/kg K.

We have $c = 2500$ J/kg K, $\theta = 16 - 14 = 2$ K, $Q = 20\,000$ J. Now

$$Q = mc\theta$$

So

$$m = \frac{Q}{c\theta} = \frac{20\,000}{2500 \times 2}$$

$$= \frac{20}{5} = 4 \text{ kg}$$

Heat exchanges

Suppose a calorimeter or other container of heat capacity C is filled with a liquid of mass m and specific heat capacity c. If the liquid rises in temperature by θ, both the liquid and the container have gained heat. The amount of heat transferred to the liquid and container is

$$Q = mc\theta + C\theta$$

since the temperature rise of the container is the same as that of the liquid.

Examples

1. 0.5 kg of water is contained in a metal vessel of heat capacity 40 J/K. An immersion heater of 100 W (100 J/s) is used for 5 min to warm the water from 20°C. What is the final water temperature? (c for water = 4200 J/kg K.)

Heat supplied, $Q = 100 \times (5 \times 60) = 30\,000$ J

If θ is the temperature rise,

heat gained by water and container $= 0.5 \times 4200 \times \theta + 40 \times \theta = 2140\theta$ J

$\therefore \quad 2140\theta = 30\,000$

$$\therefore \quad \theta = \frac{30\,000}{2140} = 14 \text{ K}$$

$\therefore \quad$ final temperature $= 20°C + 14°C$

$= 34°C$

2. In a vessel, 0.2 kg of water is raised in temperature from 10°C to 60°C in 4 min by an immersion heater. When 0.4 kg of water is placed in the same vessel and the experiment repeated, the water temperature rises from 10°C to 60°C in 7 min. Find the heat per minute supplied by the heater (c for water = 4200 J/kg K).

CHANGING FORMULAE: MECHANICS; ELECTRICITY; HEAT   97

Suppose Q is the heat per minute produced by the heater. Then

$$4Q = 0.2 \times 4200 \times (60-10) + h \qquad (1)$$

where h is the heat absorbed by the vessel during heating. Also,

$$7Q = 0.4 \times 4200 \times (60-10) + h \qquad (2)$$

since the heat absorbed by the same vessel is again h for the same temperature change from 10°C to 60°C.
Subtract (1) from (2) to eliminate h (see p. 100). Then

$$7Q - 4Q = 84\,000 - 42\,000$$
$$\therefore \quad 3Q = 42\,000$$
$$Q = \frac{42\,000}{3} = 14\,000 \text{ J/min}$$

Specific latent heat

From p. 76, to which the reader should refer, the heat Q required to evaporate a mass m of liquid of specific latent heat l *at its boiling point* is given by

$$Q = ml$$

Q is in J, m in kg and l in J/kg. In latent heat changes, the temperature remains constant.

When the vapour *condenses to liquid* at the boiling point, the heat *given up* by this change is given by $Q = ml$. If the liquid formed drops in temperature from the boiling point, more heat is given up—this time $Q = mc\theta$, where θ is the temperature change.

As an illustration, suppose 0.1 kg of steam at 100°C condenses to water whose final temperature is 40°C. If $l = 2\,200\,000$ J/kg for steam and $c = 4200$ J/kg K for water, then

$$\text{total heat given up} = ml + mc\theta$$
$$= 0.1 \times 2\,200\,000 + 0.1 \times 4200 \times (100-40)$$
$$= 220\,000 + 25\,200$$
$$= 245\,200 \text{ J}$$

Latent heat of fusion

Suppose 0.02 kg of ice at 0°C melts and forms water finally at 10°C. If the specific latent heat of fusion of ice is 330 000 J/kg, the heat needed to change the ice to water *at 0°C* is

$$ml = 0.02 \times 330\,000 = 6600 \text{ J}$$

The heat needed to raise the temperature of the water from 0°C to 10°C is

$$mc\theta = 0.02 \times 4200 \times (10-0) = 840 \text{ J}$$

So the total heat needed

$$6600 + 840 = 7440 \text{ J}$$

Example

A block of 0.1 kg of copper, $c = 400$ J/kg K, is held in boiling oxygen until it reaches this cold temperature, and then quickly transferred to water at 0°C. A crust of ice of mass 0.021 kg is now formed round the copper. Calculate the temperature of the liquid oxygen.

Suppose x°C is the liquid oxygen temperature. Then

heat gained by copper from water at 0°C

$$= mc\theta$$
$$= 0.1 \times 400 \times (0-x)$$
$$= -40x \text{ J}$$

(x is below 0°C and so has a negative value).

But heat taken from water at 0°C to form ice at 0°C

$$= ml$$
$$= 0.021 \times 330\,000 = 6930 \text{ J}$$
$$\therefore \quad -40x = 6930$$
$$x = -\frac{6930}{40} = -173$$

So liquid oxygen temperature $= -173$°C

EXERCISE 32

(Where necessary, assume c for water $= 4200$ J/kg K.)

1. A machine is driven for 10 min by a motor, and supplied with 1000 J per second of energy. The efficiency of the machine is 75%. Calculate (i) the energy 'lost' in the machine, (ii) the temperature rise of 50 kg of iron if the energy 'lost' is used to warm this mass of metal (c for iron $= 450$ J/kg K).

2. A 2 kg mass of copper at 100°C is placed inside 0.5 kg of water at 20°C inside a metal calorimeter. The water temperature rises to 35°C. Find the heat capacity of the calorimeter (c for copper $= 400$ J/kg K).

CHANGING FORMULAE: MECHANICS; ELECTRICITY; HEAT

3. An immersion heater of 100 W is placed inside an aluminium block. After 5 min, the temperature of the metal rises from 10°C to 15°C. If the heat lost to the surroundings during this time is 200 J, calculate the mass of aluminium (c for aluminium = 1000 J/kg K).

Latent heat (Where necessary assume specific latent heat of vaporization of water, $l = 2\,200\,000$ J/kg (2.2×10^6 J/kg) and specific latent heat of fusion of ice, $l = 330\,000$ J/kg (3.3×10^5 J/kg).)

4. A mass of ice at 0°C is dropped into 3.1 kg of oil at 30°C. After melting completely, the oil temperature drops to 10°C. Calculate the mass of ice (c for oil = 2000 J/kg K).

5. 0.01 kg of steam condenses to water at 30°C. The heat given up is used to raise the temperature of 0.5 kg of a metal from 10°C to 50°C. Calculate the specific heat capacity of the metal.

6. 0.05 kg of water at 15°C is cooled by a refrigerator to −5°C. Calculate the heat removed from the water, assuming the specific heat capacity of the ice is 2000 J/kg K.

9 Simultaneous and Quadratic Equations

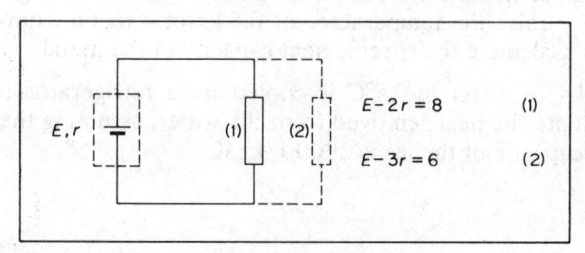

Figure 9.1.

SIMULTANEOUS EQUATIONS

Figure 9.1 illustrates a simple electric circuit with a battery of e.m.f. E and internal resistance r. When an electrical resistance (1) is connected to the battery, the theory of electric circuits shows that the relation $E - 2r = 8$ is obtained. When another resistance (2) is connected to the same battery, theory now shows that $E - 3r = 6$.

In this case, therefore, we have two unknown quantities E and r, and two equations relating them. These are examples of *simultaneous equations*. As an illustration of how to solve the equations, consider the following simultaneous equations between two unknown quantities x and y:

$$2x + 3y = 12 \qquad (1)$$
$$3x - 2y = 5 \qquad (2)$$

Elimination method

Here we get rid of one quantity, say x, leaving one equation with one unknown quantity, y. To do this, multiply both sides of equation (1) by 3. We then have

$$6x + 9y = 36 \qquad (3)$$

SIMULTANEOUS AND QUADRATIC EQUATIONS

Now multiply equation (2) by 2 on both sides. Then

$$6x - 4y = 10 \qquad (4)$$

To get rid of x, subtract (4) from (3). We have

$$+9y - (-4y) = 36 - 10$$

So

$$13y = 26$$

$$\therefore \quad y = \frac{26}{13} = 2$$

Having found the value of y, we can now substitute for y in equation (1). Then

$$2x + 6 = 12$$

$$\therefore \quad 2x = 12 - 6 = 6$$

$$x = \frac{6}{2} = 3$$

So the answer is $x = 3$ and $y = 2$.

Substitution method

We can also solve simultaneous equations by *substituting for x (or y)* from one equation into the other equation. As an example, consider again the simultaneous equations

$$2x + 3y = 12 \qquad (1)$$
$$3x - 2y = 5 \qquad (2)$$

From (1),

$$2x = 12 - 3y$$

So

$$x = \frac{12 - 3y}{2}$$

Substitute for x in equation (2). Then

$$3\left(\frac{12 - 3y}{2}\right) - 2y = 5$$

Multiply both sides of the equation by 2 to clear the fraction.

$$\therefore \quad 3(12 - 3y) - 4y = 10$$
$$\therefore \quad 36 - 9y - 4y = 10$$
$$\therefore \quad -9y - 4y = 10 - 36$$
$$-13y = -26$$
$$y = \frac{-26}{-13} = 2$$

Substituting $y = 2$ in equation (1), then

$$2x + 3 \times 2 = 12$$
$$2x + 6 = 12$$
$$2x = 12 - 6 = 6$$
$$x = \frac{6}{2} = 3$$

So the answer is $y = 2$, $x = 3$, as obtained by the elimination method.

Magnification in lenses or curved mirrors

The *linear magnification*, *m*, of an object produced by a lens or spherical mirror is the ratio *image height/object height*. *m* is related to the image distance v and the object distance u by

$$m = \frac{v}{u}$$

So if the magnification produced is 3 times, $v/u = 3$ and $v = 3u$; the image distance is 3 times the object distance. If the magnification m is $\frac{1}{2}$, image distance $= \frac{1}{2} \times$ object distance.

The relationship between u and v is also given by (p. 78):

$$\frac{1}{v} + \frac{1}{u} = \frac{1}{f}$$

where f is the focal length of the lens or mirror. So given the values of m and f, we have a pair of simultaneous equations between u and v.

Example

A converging lens of focal length 10 cm produces an upright virtual image 3 times the object height when used as a magnifying glass. Find the object distance from the lens.

Suppose x cm is the object distance, y cm is the image distance. Since $m = 3$, numerically

$$y = 3x \tag{1}$$

Using the 'real is positive' sign rule for distances, $u = +x$ cm and $v = -y$ cm. So, from the lens equation above, since $f = +10$ cm for a converging lens,

$$\frac{1}{-y} + \frac{1}{+x} = \frac{1}{+10} \tag{2}$$

From (1), we can substitute for y in (2). Then

$$\frac{1}{-3x} + \frac{1}{x} = \frac{1}{10}$$

$$\therefore \frac{-1+3}{3x} = \frac{2}{3x} = \frac{1}{10}$$

Cross-multiply,

$$3x \times 1 = 2 \times 10$$

$$\therefore x = \tfrac{20}{3} = 6\tfrac{2}{3} = 6.7 \text{ cm}$$

So

object distance = 6.7 cm

EXERCISE 33

Solve the simultaneous equations

1. $x - 2y = 2,\ 2x + y = 9$
2. $3x + 2y = 7,\ 2x - y = 0$
3. $2x - y = 5,\ x + y = 4$
4. $x + 2y = 8,\ 3x - y = 3$
5. $2x - 3y = 1,\ 3x - 4y = 3$
6. $3x + 2y = 10,\ 4x - 3y = 2$

7. A battery of e.m.f. E and internal resistance r supplies different currents when connected in turn to two different resistances (Figure 9.1). Calculate E and r from the relations below, which apply to three different batteries:
   (i) $E - 2r = 8,\quad E - 3r = 6$;
   (ii) $4E - r = 2,\quad 10E - 3r = 3$;
   (iii) $E = \tfrac{5}{3} + \tfrac{1}{3}r,\quad E = \tfrac{8}{5} + \tfrac{2}{5}r.$

8. A converging lens of focal length 5 cm in a projector produces a *real* inverted image of a slide. The linear magnification is 40. Calculate the distance of the slide from the lens. (*Note* The image distance v is *positive* as the image is real.)

9. A concave mirror, used as a shaving mirror, produces a linear magnification 2 times. If the focal length of the mirror is 20 cm, find the object distance from the mirror. (*Note* The image is *virtual*, so the image distance v is negative.)

10. An immersion heater raises the temperature of a mass m of 0.4 kg of water in a calorimeter from 10°C to 28°C in 12 min; and 0.5 kg of water in the same calorimeter from 10°C to 30°C in 16 min. Calculate the heat per minute H given out by the immersion heater and the heat capacity C of the calorimeter. (Heat gained = $H \times$ time = $(mc + C) \times$ temperature rise, where c is the specific heat capacity of water, 4200 J/kg K, p. 75.)

QUADRATIC EQUATIONS

A 'quadratic equation' is one which has the *square* of an unknown quantity, such as x^2. For example, $3x^2 = x + 1$ and $5x^2 - x - 4 = 0$ are quadratic equations.

Solving by factors

To solve a quadratic equation, first take *all* the quantities to one side of the equation, leaving zero on the other side of the equation. Then try to *factorize* the quadratic expression thus obtained. As an example, consider the equation

$$2x = 1 + \frac{1}{x}$$

To clear fractions, multiply both sides of the equation by x. Then

$$2x^2 = x + 1$$

This is a quadratic equation in x. Take all quantities to one side of the equation. So

$$2x^2 - x - 1 = 0$$

The expression $2x^2 - x - 1$ factorizes into $(2x + 1)(x - 1)$. So

$$(2x + 1)(x - 1) = 0$$

If two quantities multiplied together equal zero, then one or other of the quantities must be zero. Thus we can say now that

$$2x + 1 = 0 \quad \text{or} \quad x - 1 = 0$$

So either

$$2x = -1, \quad \text{which gives} \quad x = -\tfrac{1}{2}$$

or,

from $x - 1 = 0$, we have $x = 1$

A quadratic equation thus has two possible answers, in this case $x = -\tfrac{1}{2}$ or 1.

Example

A ball is thrown vertically upwards from the ground with a velocity of 15 m/s. Find the times taken to reach a height of 10 m ($g = 10$ m/s^2).

When the ball moves upward, it has a deceleration of 10 m/s^2, or negative acceleration of 10 m/s^2. So in

$$s = ut + \tfrac{1}{2}at^2$$

SIMULTANEOUS AND QUADRATIC EQUATIONS

$s = 10$ m, $u = 15$ m/s and $a = -10$ m/s^2. Substituting,

$$\therefore \quad 10 = 15t - \tfrac{1}{2} \times 10t^2$$
$$\therefore \quad 5t^2 - 15t + 10 = 0$$

Dividing by 5,

$$\therefore \quad t^2 - 3t + 2 = 0$$

Factorize,

$$(t-1)(t-2) = 0$$

So

$$t = 1 \text{ s} \quad \text{or} \quad 2 \text{ s}$$

Both answers are correct. The time of 1 s is the time passing the 10 m height when the ball is moving upwards, and 2 s is the *later* time when the ball again passes the 10 m height moving downwards.

Solving by formula

If the quadratic expression does not factorize, we can use a *formula* for x. Generally, if any quadratic is written as

$$ax^2 + bx + c = 0$$

where a, b and c are numbers, it can be shown by analysis that the solution is

$$x = \frac{-b + \sqrt{b^2 - 4ac}}{2a} \quad \text{or} \quad x = \frac{-b - \sqrt{b^2 - 4ac}}{2a}$$

Example

Solve $\quad 2x^2 - 3x - 4 = 0$

Comparing this equation with the general equation $ax^2 + bx + c = 0$, we see that

$a = 2$, $\quad b = -3$ (note the minus), and $\quad c = -4$ (note the minus)

So, using the formula for x,

$$x = \frac{-(-3) + \sqrt{(-3)^2 - 4 \times 2 \times (-4)}}{2 \times 2} \quad \text{or} \quad \frac{-(-3) - \sqrt{(-3)^2 - 4 \times 2 \times (-4)}}{2 \times 2}$$

$$= \frac{3 + \sqrt{9 + 32}}{4} \quad \text{or} \quad \frac{3 - \sqrt{9 + 32}}{4}$$

$$= \frac{3 + \sqrt{41}}{4} \quad \text{or} \quad \frac{3 - \sqrt{41}}{4}$$

$$= 2.4 \quad \text{or} \quad -0.9 \quad \text{(1 dec. place)}$$

EXERCISE 34

Solve the equations 1–5 by the method(s) indicated in brackets:
1. $x^2 - x - 2 = 0$ (factors).
2. $2u^2 - 3u + 1 = 0$ (factors *and* formula).
3. $3x^2 - 5x - 2 = 0$ (factors *and* formula).
4. $2x^2 + x - 4 = 0$ (formula).
5. $3a^2 - 2a - 6 = 0$ (formula).
6. A ball is thrown vertically upwards with a velocity of 25 m/s. Find the times when it passes a height of 30 m ($g = 10$ m/s^2).
7. In Question 6, the ball is thrown up with a greater velocity of 40 m/s. Find the times it now passes the height of 30 m (use formula for quadratic equation).
8. An object is placed x cm in front of a converging lens of focal length 4 cm and a real image is formed on the other side at a distance of $(x + 15)$ cm from the lens. Calculate x. (Use $1/v + 1/u = 1/f$, where $u = +x$ and $v = +(x+15)$.)
9. The electrical power in a resistor R of 2 Ω due to a current I in amperes is the same as that in a resistor of 4 Ω when the current is 1 A less than before. Calculate I (power $= I^2 R$).

10 Graphs. Linear Graphs and Applications

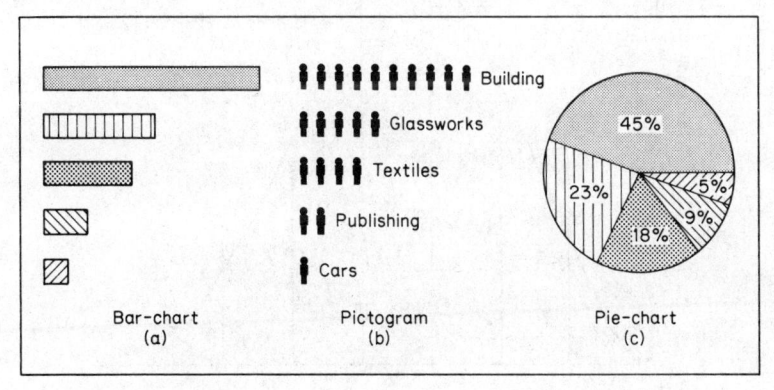

Figure 10.1 Representations of statistics.

Types of graphs

Numerical information or *statistics*, such as the attendance per week at football matches or the number of cars sold per month, can be represented by diagrams in a number of ways. This is illustrated in Figure 10.1. If a large company has 5000 employees in the building section, 2500 in glassworks, 2000 in textiles, 1000 in publishing and 500 in the cars section, the information can be shown in a *bar-chart*, Figure 10.1(a), or in a *pictogram*, Figure 10.1(b), or in a *pie-chart*, Figure 10.1(c).

Histogram; frequency curve

At the end of a particular soccer football season, the top team had played 42 matches and scored a total of 120 goals. The number of goals scored in matches varied from 0 to 6 and these were distributed among the matches as shown in the following table:

108 BASIC MATHEMATICS FOR SCIENCE

Goals	0	1	2	3	4	5	6
Matches	1	6	10	12	8	4	1

No goals were scored in 1 match, 1 goal in 6 matches, 2 goals in 10 matches, and so on. We say that 1 goal has a *frequency* 6, 2 goals a frequency 10, and so on.

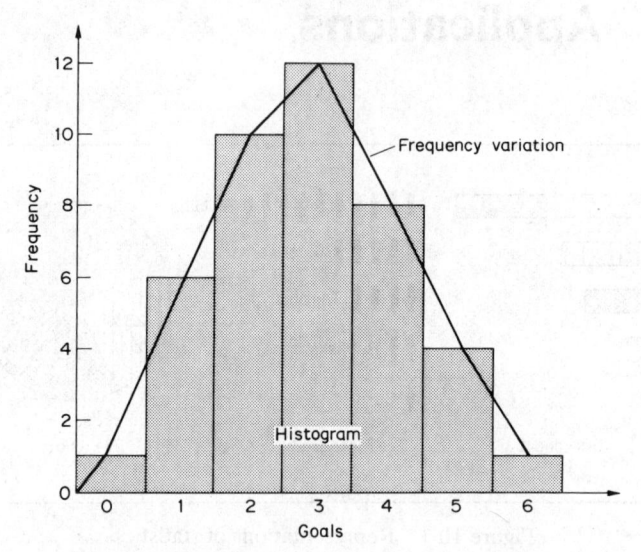

Figure 10.2 Histogram and frequency variation.

We can represent the distribution of goals among matches by plotting the number of goals along the horizontal axis and the frequency along the vertical axis. Figure 10.2 shows the result. The graph is called a *histogram*. The area of the rectangles represent the frequency and since the width is the same, the heights represent the frequencies as shown.

If the middle of the tops of the rectangles are joined, a *frequency variation* or *curve* is obtained. Figure 10.2 shows only a small number of statistics. With large numbers and narrow rectangles, a smooth curve can be drawn to show the frequency graph.

Since the total number of goals is 120 and the number of matches is 42, the *mean* or average number of goals per match is

$$\frac{120}{42} = 2.9$$

Continuous graphs

The table below shows measurements of the temperature of the air and soil near a small pond and of the water in it, from 9 a.m. to 3 p.m. on a particular day.*

	Temperature (°C)		
Time	Air	Soil	Water
9 a.m.	9.0	10.0	10.0
10	11.0	10.7	10.6
11	13.9	11.5	11.0
12 p.m.	16.1	12.3	11.4
1	16.7	13.1	11.9
2	16.4	13.3	12.3
3	15.9	13.3	12.3

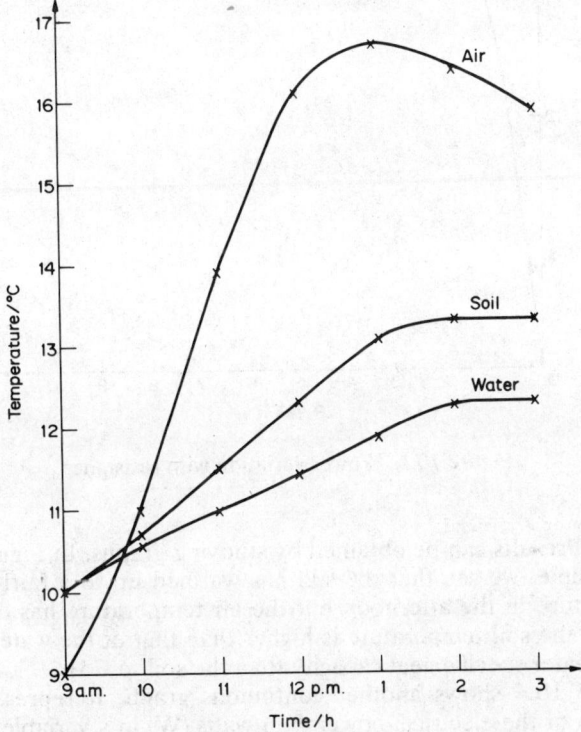

Figure 10.3 Temperature variations with time.

* Adapted from *General Science Biology* by Alan Dale (Heinemann).

Figure 10.3 shows the results when the temperature is plotted against the time. Since the temperature varies *continuously* with the time, a smooth curve is drawn through all the plotted points. Where there is no continuous variation between two quantities, for example, the number of goals and the number of matches in Figure 10.2, the points are joined by straight lines.

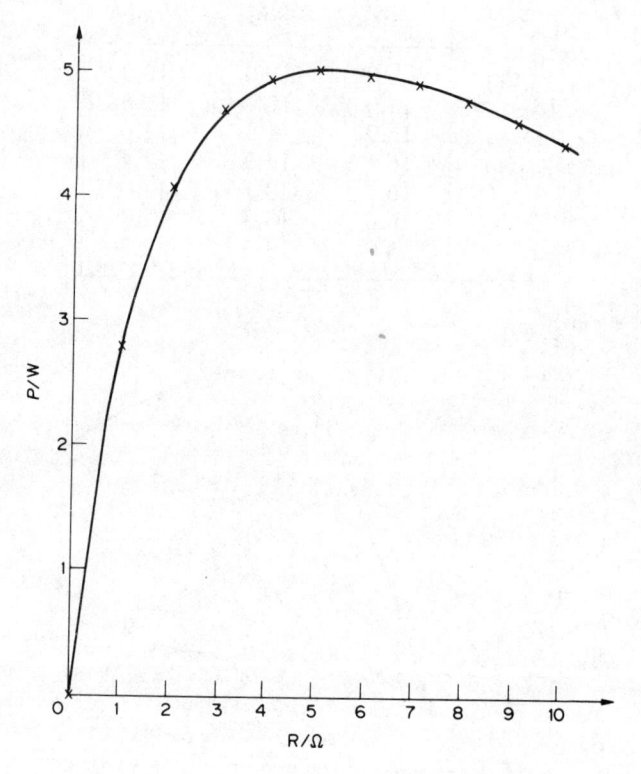

Figure 10.4   Power variation with resistance.

Useful results can be obtained by studying graphs. In Figure 10.3, for example, we see that the soil has warmed up to a fairly steady temperature in the afternoon but the air temperature has dropped. Further, the soil temperature is higher than that of the water (water has a greater specific heat capacity than the soil, p. 75).

Figure 10.4 shows another continuous graph. It represents the variation of the electrical power P in watts (W) in a variable resistor R in ohms (Ω) when the resistor is joined to a given battery. From the graph, we see that the power P is a maximum when R is 5Ω.

EXERCISE 35

1. The sales of goods by an electrical company, in thousands, varied one year as shown in the table below:

Jan	Feb	Mar	Apr	May	June	July	Aug	Sep	Oct	Nov	Dec
110	107	111	112	112	109	106	105	107	108	109	110

   (i) Draw a histogram of the monthly sales (months = horizontal axis, sales = vertical axis).
   (ii) By joining the tops of the histogram rectangles, show the variation of monthly sales.
   (iii) Which months have (a) minimum, (b) maximum sales? Can you suggest one reason for them?

2. The marks in a pre-exam test, maximum 50, were distributed among 40 pupils as follows:

Marks	1–5	6–10	11–15	16–20	21–25	26–30	31–35	36–40	41–45	46–50
No. of pupils	2	4	5	7	9	6	3	2	1	1

   (i) Draw a histogram of the marks (horizontal axis) and the number of pupils or frequency (vertical axis). Show the frequency variation by joining the tops of the histogram rectangles.
   (ii) Do you consider this a fair test? Give a reason for your answer.
   (iii) Assuming the average mark in the range 1–5 is 3, 6–10 is 8, 11–15 is 13, and so on, calculate (a) the *total marks* of all the pupils, (b) the *average or mean mark* per pupil.

3. Figure 10.5 shows the variation with temperature of the *solubility* of a crystal compound (the number of grams dissolved per litre of water). At the point A, the crystal X changes in composition to another crystal Y.
   (i) What is the temperature of maximum solubility for crystal X?
   (ii) What is the solubility at 25°C of X?
   (iii) At what temperature is the solubility of X half its maximum value?
   (iv) How does the variation of the solubility of crystal Y differ basically from that of X?
   (v) In the temperature range 20°C to 30°C, calculate the average solubility change per °C for X (solubility change per °C = solubility change ÷ temperature change).

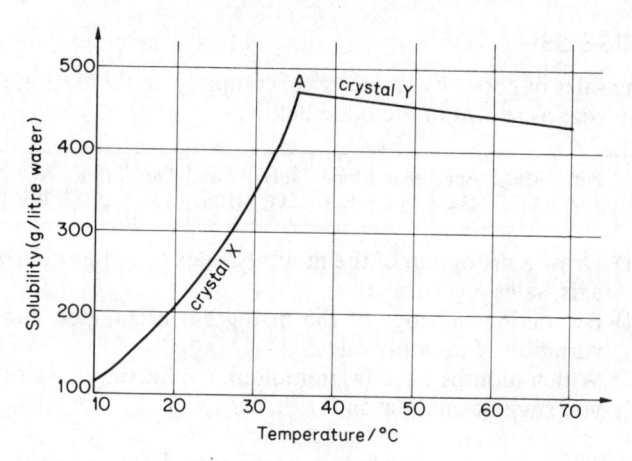

Figure 10.5.

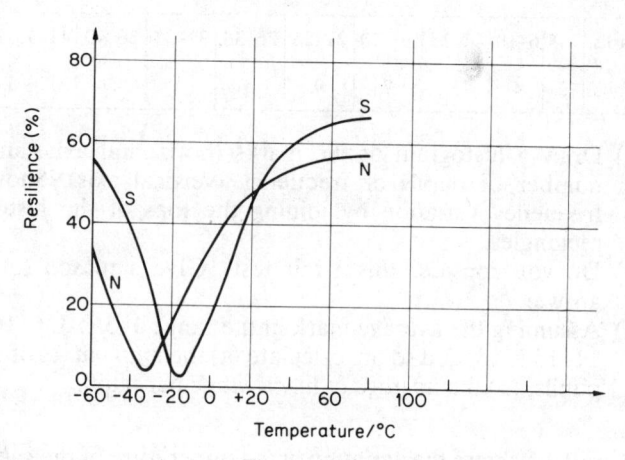

Figure 10.6.

4. Figure 10.6 shows a curve (a) for the rebound resilience of natural rubber N at various temperatures from cold to hot, and (b) for the rebound resilience of a synthetic rubber, S (rebound resilience of a rubber ball = height of rebound ÷ height dropped).
   (i) At very low temperatures, which type of rubber has a larger rebound resilience?
   (ii) At what temperature is the rebound resilience of natural rubber N a minimum?
   (iii) At what two temperatures are the rebound resiliences the same for N and S and what are the resilience values at these temperatures?

GRAPHS. LINEAR GRAPHS AND APPLICATIONS

(iv) A rubber with a low rebound resilience shows high values of rolling friction. Which rubber, N or S, may be more suitable for a car tyre tread in icy weather, when the temperature is 0°C?

5. The table below shows the height h in metres (m) dropped by a ball in a time t in seconds (s). Draw a graph of h against t (draw a smooth curve to pass through all the points).

t (s)	0	1	2	3	4	5
h (m)	0	5	20	45	80	125

From the graph, find
 (i) The time taken to drop a distance of 50 m.
 (ii) The distance dropped in 2.5 s from the start.
 (iii) The velocity at 1.5 s (velocity = gradient of tangent to curve).
 (iv) The velocity at 3.5 s. Why is this value greater than that in (iii)?

6. The table below shows values of sound intensity I when loudspeakers of various resistances R in ohms (Ω) are used in a radio receiver.

R/Ω	0	1	2	3	4	5	6	7	8	9	10
I	0	2.8	4.1	4.7	4.9	5.0	4.9	4.8	4.7	4.6	4.4

 (i) Draw a graph of I against R.
 (ii) What is the maximum sound intensity and the resistance value which produces this maximum sound?
 (iii) What resistance value would produce a sound intensity half that of the maximum?
 (iv) What resistance value would produce the same intensity as the resistance of 10 Ω?

Biology Questions

7. During an investigation into the action of salivary amylase on the digestion of starch, the following results were obtained:

Temperature (°C)	10	20	30	40	50	60	70
Rate of reaction (units per minute)	11	28	44	52	51	41	13

(i) Plot these data graphically (use zero values for the origins of both axes).
(ii) What is the optimum temperature for the reaction? (i.e. temperature giving maximum rate of reaction).
(iii) What is the rate of reaction at the optimum temperature?
(iv) Briefly account for the fall in reaction at temperatures in excess of the optimum (L).

8. With axes as shown, Graph A in Figure 10.7 shows the effect on the pulse rate of a man of five minutes exercise on a pedal machine. Graph B shows his breathing rate during the same period.
(i) Comment on the general form of the two graphs. (ii) What was the highest pulse rate recorded? (iii) What was the value of the increased pulse rate to the man? (iv) What was the highest breathing rate recorded? (v) What was the value of the increase in the breathing rate to the man? (vi) For how long after the exercise stopped was the highest pulse rate maintained? (vii) How do you account for the differences between the pulse rates at the beginning and the end of the experiment? (viii) Why did the breathing rate return to its original value more sharply and more completely than the pulse rate? (O).

9. Two sets of pea seedlings were grown from seed, one set in light and the other set in darkness. The fresh weight (i.e. the total weight, including water) was recorded at regular intervals and the following results were obtained:

Time (days)	0	5	10	15	20	25
Seedlings in light (g)	150	220	290	360	450	530
Seedlings in dark (g)	140	200	290	400	530	600

Time (days)	30	35	40	45	50
Seedlings in light (g)	610	710	840	1100	1300
Seedlings in dark (g)	740	710	600	420	280

(i) Plot these data graphically using 'fresh weight' as vertical axis and 'time' as horizontal axis.
(ii) Write a short paragraph to explain these results (L).

10. Some flower seeds were stored at a constant temperature of 8°C in two separate rooms. The batch in the first room was maintained at 30% relative humidity and the second batch at 80% relative humidity. Samples were taken at intervals and were germinated under identical conditions. The percentage germination was recorded as follows:

GRAPHS. LINEAR GRAPHS AND APPLICATIONS   115

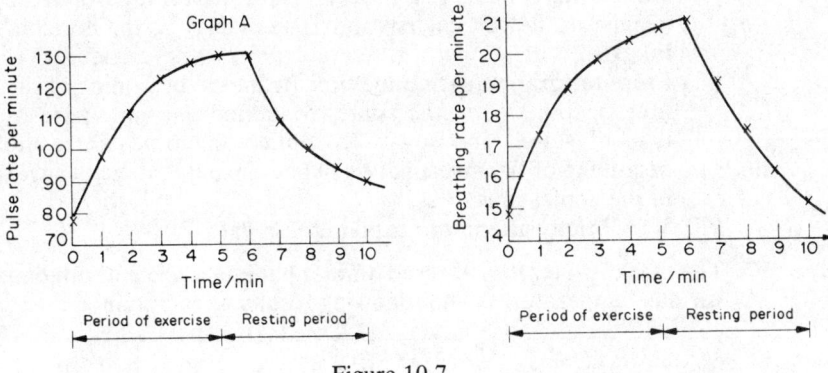

Figure 10.7.

Years of storage	0.5	1	2	3	4
Percentage germination at 30% humidity	55.5	52.0	47.0	43.5	42.0
Percentage germination at 80% humidity	50.0	43.0	31.5	26.0	22.0

(i) Plot these data together graphically, using 'percentage germination' as the vertical axis and 'years of storage' as the horizontal axis.
(ii) Obtain from the graph: (a) the percentage germination of seeds stored for 2.5 years at 80% relative humidity, (b) the maximum time that seeds could be stored at 30% relative humidity to ensure a germination rate of 50%.
(iii) What can be deduced *from the graph alone* about the storage of seeds for germination? (*L*).

11. A sample of keeping apples were analysed at regular intervals from about three weeks after the beginning of their development on the tree until the end of some weeks of storage. At each analysis, the amount of starch and sugar (sucrose) was recorded as a percentage of the fresh weight of the apples. The following data were obtained:

Days	0	25	50	75	100
Starch (% fresh weight)	0.45	0.95	1.35	1.20	0.50
Sugar (% fresh weight)	0.30	0.65	1.60	2.10	2.50

Days	125	150	175	200
Starch (% fresh weight)	0.00	0.00	0.00	0.00
Sugar (% fresh weight)	2.75	2.65	2.40	2.20

(i) Plot these data, graphically, using 'percentage of fresh weight' as the vertical axis and 'time in days' as the horizontal axis.
(ii) From the graph, (a) find the time (from the beginning of the investigation) when the apples contained the same percentage of starch and sugar, (b) obtain the time (from the beginning of the investigation) when the percentage of sugar in the apples falls to 2.5%.
(iii) Very briefly attempt to explain these data (*L*).

12. The rate of water loss of three different animals was measured at various temperatures. The following results were obtained:

Temperature (°C)	10	20	30	40	50	60
Rate of water loss (mg/cm² h)						
Woodlouse	6.0	7.5	10.0	13.0	17.5	20.0
Spider	1.0	1.5	2.0	4.0	10.0	19.0
Scorpion	1.0	1.5	2.0	2.5	3.0	9.5

(i) Plot these data graphically, using 'rate of water loss' as the vertical axis and 'temperature' as the horizontal axis.
(ii) Find out from the curves, the *difference* in the rate of water loss: (a) between woodlouse and spider at 45°C, (b) between woodlouse and scorpion at 55°C.
(iii) State which of these organisms seems at least partly able to control water loss and briefly give reasons for your answer (*L*).

LINEAR GRAPHS AND APPLICATIONS IN PHYSICS

Linear or straight-line graphs are particularly useful in physics. In this section we shall begin with a general discussion of linear graphs and then deal with their applications.

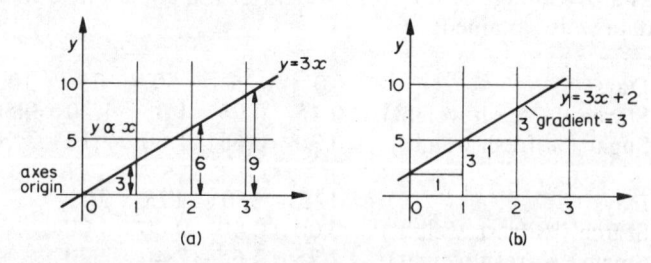

Figure 10.8.

Linear graphs passing through origin

An equation of the form $y = 3x$ is a *first degree* equation in x or a *linear function* of x.

When $x = 0$, then, from $y = 3x$, we obtain $y = 0$. So the graph of $y = 3x$ passes through the origin. From the equation, when $x = 1$, then $y = 3$; so when x changes by 1 unit from 0, y changes by 3 units. When $x = 2$, then, from the equation, $y = 6$. Once again, a change in x of 1 unit produces a change of 3 units in y.

From Figure 10.8(a) we now see that the *gradient* of the graph $y = 3x$ is constant and equal to 3. So the graph is a *straight line passing through the origin* and has a gradient 3.

We can see from $y = 3x$ that if we double the value of x, then the value of y becomes double. If we treble the value of x, then the value of y is trebled. We say that

$$y \text{ is directly proportional to } x$$

or

$$y \propto x$$

All graphs of the form $y = kx$, where k is a number, are 'directly proportional' graphs. And just as the '3' in $y = 3x$ represents the gradient of the graph, so the number 'k' represents the gradient of the graph $y = kx$.

Linear graphs not passing through origin

Consider the equation $y = 3x + 2$. This is a first degree equation in x. When $x = 0$, then, from the equation, $y = 2$. So this graph does *not* pass through the origin.

When $x = 1$, then, from $y = 3x + 2$, $y = 5$. So the gradient of the graph between $x = 0$ and $x = 1$ is

$$\frac{\text{change in } y}{\text{change in } x} = \frac{5-2}{2-1} = 3$$

When $x = 2$, then, from the equation, $y = 8$. So the gradient from $x = 1$ to $x = 2$ is

$$\frac{\text{change in } y}{\text{change in } x} = \frac{8-5}{3-2} = 3$$

The gradient is therefore constant. The graph of $y = 3x + 2$, in fact, is a straight line (*see* Figure 10.8(b)). It does not pass through the origin.

Positive and negative gradients

As we have seen, the graph $y = 3x + 2$ slopes upwards from left to right (Figure 10.9(a)). The line has a 'positive' gradient of value 3.

The graph $y = 5 - 2x$, however, is a straight line graph which slopes downwards from right to left (Figure 10.9(b)). This time the gradient is 'negative' and equal to -2.

It may be noted that the number multiplying x in the equation for y, called the 'coefficient' of x, is equal to the gradient.

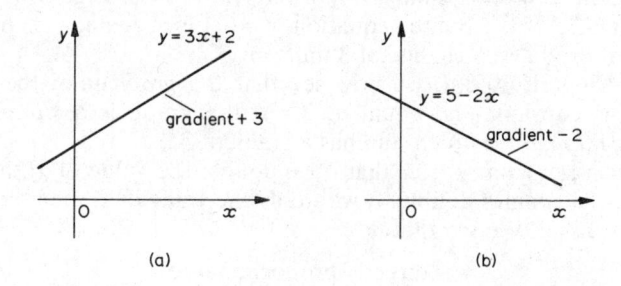

Figure 10.9.

EXERCISE 36

Using three values of x and y, plot the following linear graphs. For each graph, (i) find its gradient, (ii) state if $y \propto x$.

1. $y = 4x + 2$
2. $y = 2x$
3. $y = 5 - 3x$
4. $y = -4x$
5. $2y = 3x - 2$

Some linear graphs in physics

In Science, many relationships between two quantities have been found by making measurements of them, and then plotting one quantity against the other. If a *straight line* graph can be drawn through the plotted points, then we can write down a general relationship between the quantities as we soon show.

In this section we consider linear graphs found in different branches of physics. The principles also apply to chemistry and biology.

Elasticity: Hooke's Law

To investigate the elasticity of a spiral spring, we suspend the spring from one end and load the other end with increasing weights (Figure 10.10(a)). As each weight is added, the extension x of the spring from its original length is measured using a ruler.

GRAPHS. LINEAR GRAPHS AND APPLICATIONS

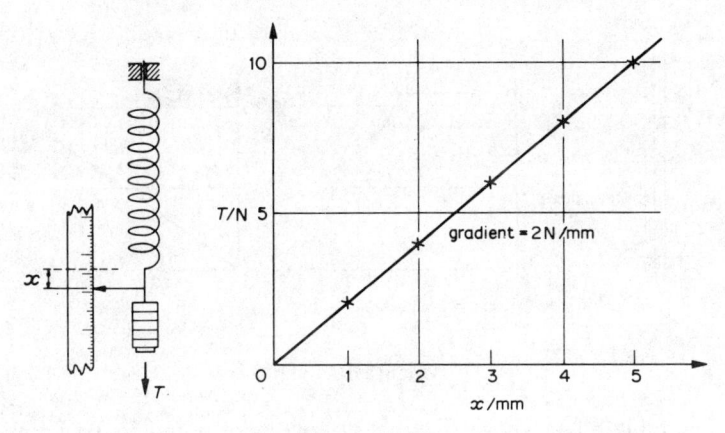

Figure 10.10.

The results are shown in the table below. The tension (force) T which stretches the spring is equal to the weight and the values of T vary from 0 to 5 N.

T (N)	0	1	2	3	4	5
x (mm)	0	2	4	6	8	10

Figure 10.10(b) shows the graph obtained when T is plotted against x. The points lie on a *straight line passing through the origin.* So we deduce that

$$T \propto x$$

or

$$T = kx$$

where k is a constant. The law *extension directly proportional to tension* was discovered by Hooke, and is known as *Hooke's Law*. It is true up to the 'elastic limit' of the spring.

From Figure 10.10(b), we see that the gradient in this case is 2 N/mm. So we can write the relationship between T and x for this spring as

$$T = 2x$$

when T is in newtons and x is in millimetres.

Gases: Boyle's Law

By means of a pump, the pressure p of a fixed mass of gas such as air can be varied, and the volume V of the gas measured each time (Figure 10.11(a)).

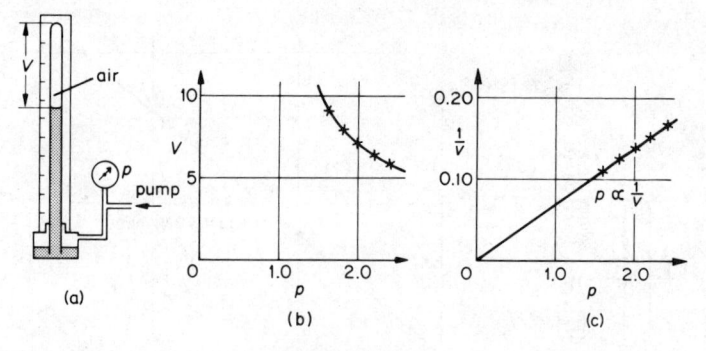

Figure 10.11.

The table below shows five measurements of p and V. To simplify the arithmetic, we record the values as so many 'units'.

p	1.6	1.8	2.0	2.2	2.4
V	9.0	8.0	7.2	6.5	6.0
$1/V$	0.11	0.13	0.14	0.15	0.17

Figure 10.11(a) shows the graph obtained when V is plotted against p. The points lie on a smooth curve. This shows that there is a certain relationship between p and V, but this can not be easily deduced.

To see if a straight line graph can be obtained, we calculate the values of $1/V$, the reciprocal of V. The results for $1/V$ are shown in the table. When $1/V$ is plotted against p, this time the points are found to lie on a straight line. Figure 10.11(b).

We therefore deduce that

$$\frac{1}{V} \propto p$$

or

$$\frac{1}{V} = kp$$

where k is a constant. So, cross-multiplying, $1 = kpV$

$$\therefore \quad pV = \frac{1}{k} = \text{constant}$$

The relationship $pV = constant$ for a mass of gas at constant temperature is known as Boyle's Law, after the discoverer. This is the

GRAPHS. LINEAR GRAPHS AND APPLICATIONS   121

equation between p and V for the curved graph shown in Figure 10.4(a), but we were unable to deduce it until we found the linear relationship between $1/V$ and p.

Electrical resistance

The electrical resistance R of a length of Nichrome wire can be measured by a voltmeter–ammeter method, using the circuit shown in Figure 10.12(a). When the p.d. V is varied from 0 to 5 V in steps of 1 V, the current I flowing in the wire has corresponding values shown in the table below.

V (V)	0	1	2	3	4	5
I (A)	0	0.2	0.4	0.6	0.8	1.0

When V is plotted against I, the points are found to lie on a straight line passing through the origin (Figure 10.12(b)). Now $V = RI$, or $R = V/I$ = gradient of line. From the graph in Figure 10.12(b)

$$\text{gradient} = \frac{4 \text{ V}}{0.8 \text{ A}} = 5.0 \ \Omega$$

So the resistance R of the wire is 5.0 Ω.

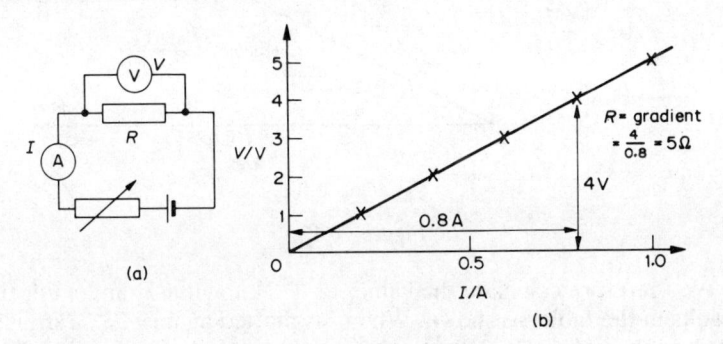

Figure 10.12.

The gradient method gives a more accurate value of R than calculating R from individual values of V and I.

Pendulum method for g

A value for g, the acceleration due to gravity, can be found by varying the length l of a simple pendulum, and measuring each time the period of oscillation, T (Figure 10.13(a)). Some values of l and T are shown in the table below.

l (m)	0.8	1.0	1.2	1.4	1.6
T (s)	1.80	2.02	2.21	2.40	2.54
T^2 (s^2)	3.2	4.0	4.8	5.8	6.4

Now theory shows that

$$T = 2\pi\sqrt{\frac{l}{g}}$$

Squaring both sides of the equation and re-arranging, we find that

$$g = 4\pi^2 \frac{l}{T^2} \qquad (1)$$

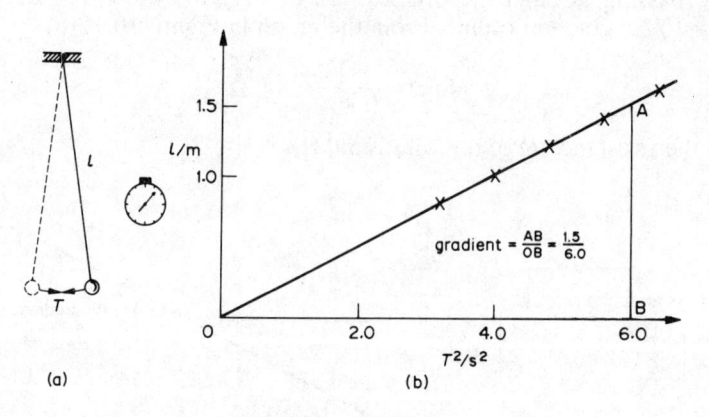

Figure 10.13.

We therefore calculate the values of T^2, knowing T, and enter the results in the table, as shown. When l is plotted against T^2, a straight line graph, practically passing through the origin, is obtained. The gradient of this line is l/T^2. From Figure 10.13(b)

$$\text{gradient} = \frac{AB}{OB} = \frac{l}{T^2} = \frac{1.5}{6.0}$$

Hence, from (1),

$$g = 4\pi^2 \times \frac{AB}{OB} = 4\pi^2 \times \frac{1.5}{6.0} = 9.9 \text{ m/s}^2$$

So g can be calculated.

GRAPHS. LINEAR GRAPHS AND APPLICATIONS 123

EXERCISE 37

1. The following values of V and I were measured for a wire in a determination of its resistance:

V (V)	1	2	3	4	5
I (A)	0.1	0.2	0.3	0.4	0.5

(i) Plot V against I. (ii) From the graph, find the value of the resistance, R.

2. In an experiment on the stretching of an elastic spring, the following measurements of extension x and tension T were recorded, the elastic limit not being exceeded:

T (N)	2	4	6	8	10
x (mm)	0.5	1.0	1.5	2.0	2.5

(i) Plot T against x. (ii) From the graph, find the value of k for the spring if $T = kx$.

3. In an experiment, the acceleration a of a trolley was measured for different value of force, F. Results were as follows:

F (N)	1	2	3	4	5
a (m/s^2)	0.5	1.0	1.5	2.0	2.5

(i) Plot F against a. (ii) State your deduction from the graph you draw passing through the plotted points. (iii) Calculate the mass m of the trolley from $F = ma$.

4. A steel ball falls through various distances s starting from rest, and the times t are measured. Results were as follows:

s (m)	0	0.5	0.7	0.9	1.1
t (s)	0	0.32	0.38	0.43	0.47
t^2 (s^2)	0	0.10	0.14	0.18	0.22

(i) Plot a graph of s against t. (ii) Plot a graph of s against t^2. (iii) If $s = \frac{1}{2}gt^2$, use one of the graphs to calculate a value of g from its gradient.

5. The heat Q produced in 5 min in a wire of constant resistance is measured for different values of current, I. The results are shown in the table.

Q (J)	3000	12 000	27 000	48 000	75 000
I (A)	0.5	1.0	1.5	2.0	2.5
I^2 (A^2)	0.25	1.0	2.25	4.0	6.25

(i) Plot a graph of Q against I. (ii) Plot a graph of Q against I^2. (iii) From your graphs, write down a conclusion about the relationship between the heat and the current.

6. The resistance R of the same length of a given conductor is measured for different values of A, its cross-sectional area. The results are shown in the table below.

R (Ω)	4.8	2.4	1.6	1.2	1.0
A (mm^2)	0.5	1.0	1.5	2.0	2.5
$1/A$	2.0	1.0	0.67	0.5	0.4

(i) Plot R against A. (ii) Plot R against $1/A$. (iii) From the graphs, write down your deduction.

7. The illumination E of a surface is measured at different distances r from a small lamp and the results are shown in the table below.

E (lux)	24.0	10.7	6.0	3.8	2.7
r (m)	0.10	0.15	0.20	0.25	0.30
$1/r^2$	100	44.4	25.0	16.0	11.1

(i) Plot E against r. (ii) Plot E against $1/r^2$. (iii) From the graphs, write down the relationship between E and r.

8. The power P produced in a constant resistance at different voltages V is measured and the results are given in the table.

V (V)	10	20	30	40	50
P (W)	2	8	18	32	50

By plotting a suitable graph, find the relationship between P and V.

11 Proportional Relationships

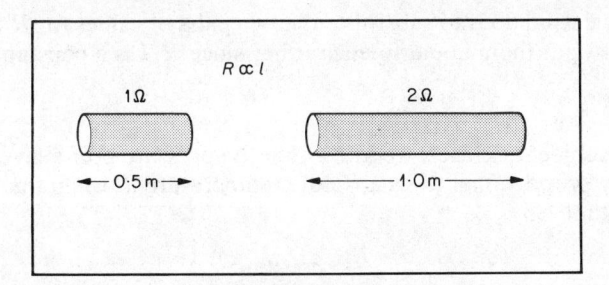

Figure 11.1.

In the previous chapter we saw that the resistance R of a wire of given diameter was *directly proportional* to its length l. We write this statement

$$R \propto l$$

or

$$\text{the ratio } \frac{R}{l} = \text{constant}$$

These relationships mean that when the length l is doubled, the resistance R is doubled; when l is halved, then R is halved.

Generally, if one quantity y is directly proportional to another quantity x, then

$$\frac{y}{x} = \text{constant}$$

Resistance

Suppose a wire of given diameter has a resistance of 15.0 Ω when its length is 3 m and the length l is required when the resistance is 4.5 Ω. Then, since

$$\frac{R}{l} = \text{constant},$$

then

$$\frac{15.0}{3} = \frac{4.5}{l}$$

Cross-multiply,

$$15.0 \times l = 4.5 \times 3$$

$$l = \frac{4.5 \times 3}{15.0} = 0.9 \text{ m}$$

All we need do is to substitute the two pairs of values for R and l in R/l, and put them equal to each other since R/l is a constant value.

Gases

In the subject of Gases, we know that the pressure p of a given gas is directly proportional to its absolute temperature T when the volume is constant. So

$$\frac{p}{T} = \text{constant}$$

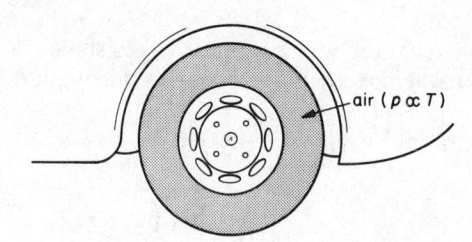

Figure 11.2.

Suppose a car tyre has a pressure of 2.0×10^5 N/m² at an absolute temperature of 280 K, and the pressure p at an absolute temperature 290 K is required (Figure 11.2). Then substituting the pairs of values in p/T, we have

$$\frac{2.0 \times 10^5}{280} = \frac{p}{290}$$

Cross-multiply,

$$2.0 \times 10^5 \times 290 = p \times 280$$

So

$$p = \frac{2.0 \times 10^5 \times 290}{280}$$

$$= 2.1 \times 10^5 \text{ N/m}^2$$

Changing gas volumes to s.t.p.

A general relationship between pressure p, volume V, and absolute or kelvin temperature T for a given mass of gas is

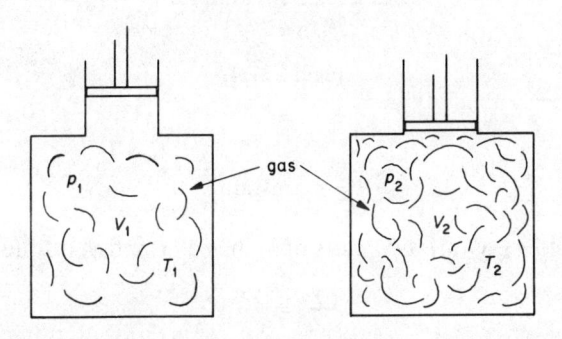

$$\frac{pV}{T} = \text{constant}$$

Figure 11.3.

So if p_1, V_1, T_1 and p_2, V_2, T_2, are two sets of values for the gas, then (Figure 11.3)

$$\frac{p_1 V_1}{T_1} = \frac{p_2 V_2}{T_2}$$

Suppose a volume of 45.0 cm^3 of oxygen is collected in a chemical experiment at a pressure of 768 mmHg and temperature of 15°C or 288 K. Standard temperature and pressure (s.t.p.) is 0°C (273 K) and 760 mmHg. To change the volume to s.t.p. we write, from above,

$$\frac{760 \times V}{273} = \frac{768 \times 45.0}{288}$$

So

$$V = \frac{768 \times 45.0 \times 273}{288 \times 760}$$
$$= 43.1 \text{ cm}^3$$

Square-law relationships

If the strength x of a magnet is doubled, the force of attraction y on iron increases four times (Figure 11.4). If x increases 3 times, then y increases 9 times. So y is proportional to the *square* of x, or

$$y \propto x^2$$

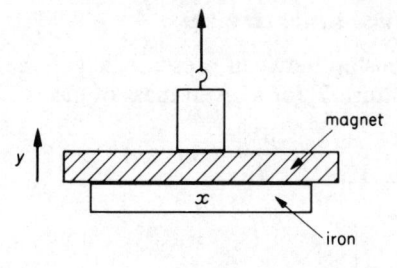

Figure 11.4.

Thus

$$\frac{y}{x^2} = \text{constant}$$

If x_1, y_1 and x_2, y_2 are two pairs of values of x and y, it follows that

$$\frac{y_1}{x_1^2} = \frac{y_2}{x_2^2}$$

Electrical heating

The heat H produced in a given time by a current I in a wire of constant resistance is proportional to the *square* of the current. In this case, then (Figure 11.5):

$$H \propto I^2$$

or

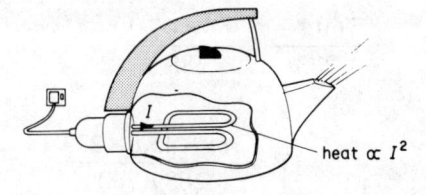

$$\frac{H}{I^2} = \text{constant}$$

Figure 11.5.

Suppose a current of 2.0 A produces 4000 J of heat in a certain time in a resistance wire and a current I is needed to produce a smaller amount of heat 1600 J in the same time. Then, since H/I^2 is the same for both pairs of values of heat and current,

$$\frac{4000}{2.0^2} = \frac{1600}{I^2}$$

Cross-multiplying,
$$4000 \times I^2 = 1600 \times 2.0^2 = 6400$$
$$\therefore \quad I^2 = \frac{6400}{4000} = 1.6$$
$$\therefore \quad I = \sqrt{1.6} = 1.3 \text{ A}$$

EXERCISE 38

1. The extension x of a spiral spring is proportional to the tension T in the spring when the elastic limit is not exceeded. If x is 1.20 mm when T is 25 N, find (i) the extension when T is 20 N, (ii) the tension when x is 1.50 mm.

2. A cube X has a side whose length is twice that of another cube Y. What is the ratio of (i) the surface area of X to that of Y, (ii) the volume of X to that of Y?

3. The resistance of a given wire is proportional to its length. The resistance of a length 1.2 m is 4.8 Ω. Find (i) the resistance of a length 1.0 m, (ii) the length with a resistance 3.6 Ω.

4. The pressure p of a gas at constant volume is proportional to its kelvin or absolute temperature T. If the pressure is 1.0×10^5 N/m² at 273 K, find (i) the pressure at 290 K, (ii) the temperature when the pressure is 1.2×10^5 N/m².

5. The pressure p, volume V, and kelvin temperature T of a given mass of matter are related by $pV/T =$ constant. Find (i) the volume at 0°C and 760 mmHg of 56.4 cm³ of oxygen collected at 15°C and 770 mmHg, (ii) the volume at 0°C and 760 mmHg of 120.0 cm³ of oxygen collected at 17°C and 750 mmHg.

6. The heat produced in a wire of constant resistance is proportional to the square of the current.
   If 8000 J of heat is produced by a current of 0.5 A, find (i) the heat produced by a current of 1.5 A, (ii) the current which produces 32 000 J of heat.

7. With a stretched string of constant length, the frequency of the note produced is proportional to the square-root of the tension. If the frequency is 800 Hz for a tension of 20.0 N, calculate (i) the frequency for a tension of 80.0 N, (ii) the tension when the frequency is 160 Hz.

8. The kinetic energy of a moving object of constant mass is proportional to the square of its velocity. If the kinetic energy is 500 J when the velocity is 10 m/s, find (i) the velocity when the kinetic energy is 2000 J, (ii) the kinetic energy when the velocity is 2 m/s.

9. The period of a simple pendulum is proportional to the square root of its length. If the period is 2.0 s for a length of 1.00 m, calculate (i) the length which gives a period of 1.5 s, (ii) the period for a length of 2.50 m.

10. If a p.d. of 2.25 V produces a current of 0.9 A in a given resistor, (i) what current will 1.75 V produce, (ii) what p.d. is required to produce a current of 0.4 A? (Current is proportional to p.d.)

Inversely-proportional relationships

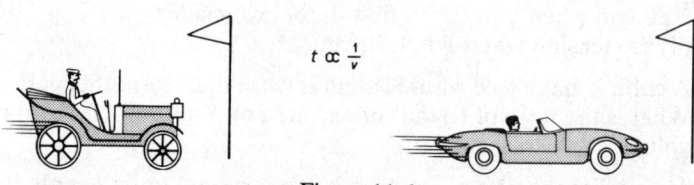

Figure 11.6.

If a car travels faster between two towns, the time taken becomes shorter (Figure 11.6). Thus if the speed is doubled, the time is halved. Conversely, if the speed is halved, the time is twice as long. We say that the time t is *inversely-proportional* to the speed v. We write this statement as

$$t \propto \frac{1}{v}$$

and $1/v$ is called the 'reciprocal' of v. So if two pairs of values for time and speed are t_1, v_1 and t_2, v_2, then

$$\frac{t_1}{t_2} = \frac{1/v_1}{1/v_2} = \frac{v_2}{v_1}$$

Cross-multiplying, we see that, for an inversely-proportional relationship,

$$t_1 v_1 = t_2 v_2, \quad \text{or} \quad vt = \text{constant}$$

Boyle's Law for gases

Boyle's Law for gases states that at a constant temperature, the volume V of a given gas is inversely-proportional to its pressure p. So, from above,

$$pV = \text{constant}$$

or

$$p_1 V_1 = p_2 V_2$$

PROPORTIONAL RELATIONSHIPS

where p_1, V_1 and p_2, V_2 are two sets of values of pressure and volume.

In Figure 11.7, a gas at constant temperature has a pressure p and a volume V. When the pressure is doubled to $2p$, the volume is halved to $V/2$, since *pressure × volume* is a constant value.

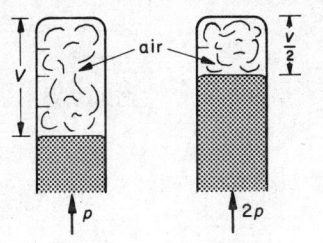

Figure 11.7.

Suppose a gas has a volume of 300 cm^3 at a pressure of $1.0 \times 10^5 \text{ N/m}^2$. If the volume is V at a new pressure of $1.5 \times 10^5 \text{ N/m}^2$ when the temperature is constant, then, from $pV = $ constant,

$$1.5 \times 10^5 \times V = 1.0 \times 10^5 \times 300$$
$$\therefore V = \frac{1.0 \times 10^5 \times 300}{1.5 \times 10^5}$$
$$= 200 \text{ cm}^3$$

If the gas pressure p is now decreased and the volume of the gas expands to 500 cm^3 at constant temperature, then, from $pV = $ constant,

$$p \times 500 = 1.5 \times 10^5 \times 200$$
$$\therefore p = \frac{1.5 \times 10^5 \times 200}{500}$$
$$= 0.6 \times 10^5 \text{ N/m}^2$$

Wavelength and frequency of waves

The higher the frequency f of a wave, the shorter is its wavelength λ. The relationship is an inversely-proportional one. So

$$f\lambda = \text{constant}$$

where the constant is the *speed* of the wave. For electromagnetic waves such as radio waves or light waves, the speed in a vacuum is about 300 million metres per second or 3×10^8 m/s (Figures 11.8(a) and (b)).

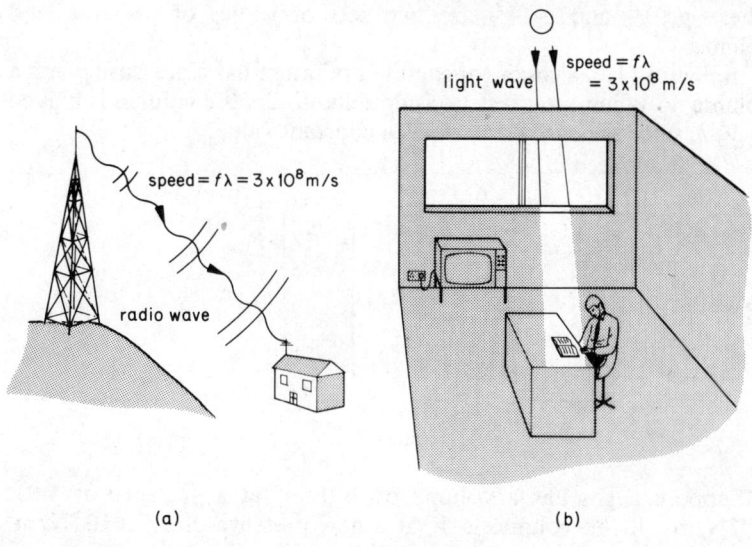

Figure 11.8.

The BBC 2 radio transmitter sends out waves of wavelength 1500 m and frequency 2.0×10^5 Hz. The BBC 1 transmitter sends out radio waves of wavelength 247 m at a frequency f say. Since $f\lambda$ = constant, then, for the two pairs of values,

$$f \times 247 = 2 \times 10^5 \times 1500$$

$$\therefore f = \frac{2 \times 10^5 \times 1500}{247}$$

$$= 12.1 \times 10^5 \text{ Hz}$$

A particular yellow light has a frequency of 5×10^{14} Hz. Since $f\lambda$ is the same for light waves as radio waves, the wavelength λ of this light wave is given by

$$2 \times 10^5 \times 1500 = 5 \times 10^{14} \times \lambda$$

$$\therefore \lambda = \frac{2 \times 10^5 \times 1500}{5 \times 10^{14}} = \frac{2 \times 5}{5} \times 10^{5+2-14}$$

$$= 6 \times 10^{-7} \text{ m}$$

Inverse-square relationships

If the distance x from a small lamp is doubled, the intensity of the light y decreases to one-quarter of its original value (Figure 11.9(a)). If x is trebled, y decreases to one-ninth. So y is inversely-proportional to the *square* of x. This is written

$$y \propto \frac{1}{x^2}$$

So

$$y = \text{constant} \times \frac{1}{x^2}$$

or

$$yx^2 = \text{constant}$$

This means that if x_1, y_1 and x_2, y_2 are two pairs of values of x and y, then, for an inverse-square law relationship,

$$y_1 x_1^2 = y_2 x_2^2$$

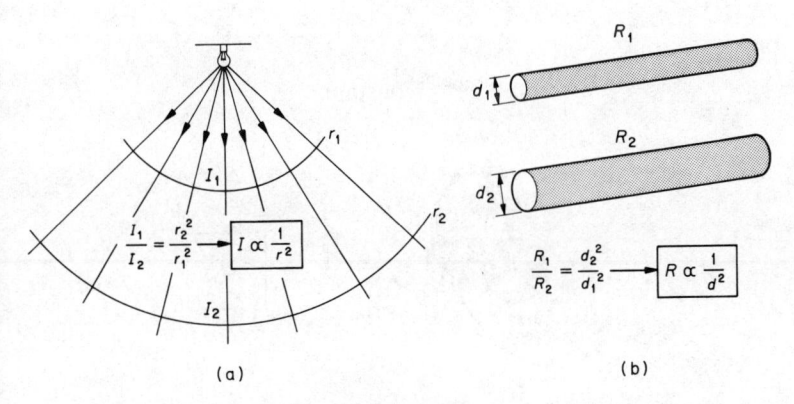

Figure 11.9.

Resistance and diameter of cross-section

The electrical resistance R of a given length of uniform wire (one of constant thickness) is inversely-proportional to the square of its diameter d (Figure 11.9(b)). So

$$R \propto \frac{1}{d^2}$$

or

$$Rd^2 = \text{constant}$$

Suppose a uniform wire of diameter 0.50 mm has a resistance of 9 Ω, and we need to find the resistance of a wire of the same material and length which has a diameter of 0.30 mm. Then, since Rd^2 is the same for the two pairs values,

$$R \times 0.30^2 = 9 \times 0.50^2$$

$$\therefore R = \frac{9 \times 0.50^2}{0.30^2}$$

$$= \frac{9 \times 0.25}{0.09} = 25 \, \Omega$$

Law of gravitation

Newton found that the force of attraction F between two masses is inversely-proportional to the square of their distance r apart. So

$$F \propto \frac{1}{r^2}$$

or

$$Fr^2 = \text{constant}$$

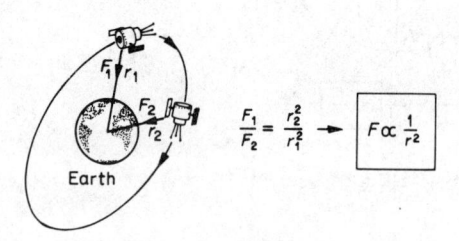

Figure 11.10.

Suppose the force of attraction by the Earth on a satellite is 9000 N when the satellite is 10 000 m from the centre of the Earth. We can find where the force of attraction will reduce to 4000 N by using $Fr^2 = $ constant for the two pairs of values of F and r. Then

$$4000 \times r^2 = 9000 \times 10\,000^2$$

$$\therefore r = \sqrt{\frac{9000 \times 10\,000^2}{4000}}$$

$$= \sqrt{\frac{9 \times 10\,000^2}{4}} = 15\,000 \text{ m}$$

So the force reduces to 4000 N at a distance of 15 000 m from the centre of the Earth.

EXERCISE 39

1. The frequency of a note produced by a string of constant tension is inversely-proportional to its length. The frequency is 450 Hz for a length 60.0 cm. Find (i) the frequency for a length 54.0 cm, (ii) the length which produces a frequency of 1800 Hz.

2. The resistance of a wire of constant length and material is inversely-proportional to the square of its diameter. If the resistance is 10.0 Ω for a diameter of 0.5 mm, calculate (i) the resistance for a diameter of 1.0 mm, (ii) the diameter which produces a resistance of 250 Ω.

3. The volume of a given gas at constant temperature is inversely-proportional to the pressure. If the volume is 150 cm^3 at 1.2×10^5 N/m^2 pressure, find (i) the volume at 1.5×10^5 N/m^2; (ii) the pressure when the volume is 100 cm^3.

4. The force of attraction on a given mass by the Earth is inversely-proportional to the square of its distance from the centre of the Earth. The force is 10.0 N when this distance is 6400 km. Calculate (i) the force when the distance is 8000 km, (ii) the distance when the force is 2.5 N.

5. The power produced in a resistor by a constant voltage is inversely-proportional to the resistance. If the power is 60 W for a resistance of 520 Ω, find (i) the power for a resistance of 600 Ω, (ii) the resistance which produces a power of 80 W.

6. The resistance R of a wire is directly proportional to its length l and inversely-proportional to its cross-sectional area A. (i) Write down an equation relating R to l and A. (ii) If R is 8 Ω when l is 1.2 m and A is 1.0 mm^2, calculate the resistance when l is 1.8 m and A is 0.5 mm^2. (iii) If R is 10 Ω and A is 1.5 mm^2, calculate l.

7. The power P produced in a conductor is proportional to the square of the applied voltage V and inversely-proportional to its length l. (i) Write down an equation relating P to V and l. (ii) If P is 40 W when V is 12 V and l is 3.6 m, calculate the length when the power produced by 6 V is 60 W. (iii) Find V if the power is 150 W when l is 1.5 m.

8. When a constant p.d. is applied to a resistor whose temperature is unaltered, the current flowing is inversely-proportional to the resistance. If a current of 0.5 A flows when the resistance is 4.5 Ω, (i) what current flows when the resistance is 7.5 Ω, (ii) what resistance produces a current of 2.5 Ω?

12 Chemistry Calculations: Reactions; Electrolysis

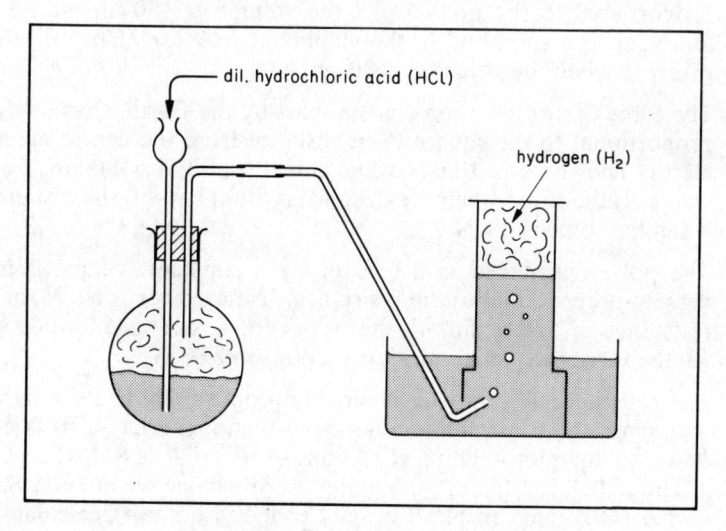

Figure 12.1.

In this section we shall deal with some chemical calculations which concern reactions and electrolysis.

MASSES AND VOLUMES IN REACTIONS

Masses

1. Consider the reaction between zinc (Zn) and excess of dilute hydrochloric acid (HCl), which produces hydrogen gas (H_2) (Figure 12.1). The equation is

$$Zn + 2HCl \rightarrow ZnCl_2 + H_2$$

CHEMISTRY CALCULATIONS: REACTIONS; ELECTROLYSIS 137

The relative atomic masses are: Zn = 65, Cl = 35.5, H = 1. So

65 g of zinc yield 2 g of hydrogen

If 5.0 g of zinc is used, suppose x g of hydrogen is produced. Then, since

mass of hydrogen $\propto$ mass of zinc

we can write, from this proportional relation,

$$\frac{x}{2} = \frac{5}{65}$$

$$\therefore \quad x = \frac{5}{65} \times 2 = 0.15 \text{ g}$$

Note 5 g of zinc produces proportionately *less* hydrogen than 65 g. So the fraction is $\frac{5}{65}$ of 2 g (*not* $\frac{65}{5}$ of 2 g).

2. From the same chemical reaction, namely,

$$Zn + 2HCl \rightarrow ZnCl_2 + H_2$$

we can calculate the mass of hydrogen chloride HCl in solution which would be needed to dissolve completely 5.0 g of zinc. Since the relative molecular mass of HCl is

$$1 + 35.5 = 36.5$$

then, from the reaction,

2 × 36.5 g of hydrogen chloride dissolve 65 g of zinc

or

73 g of hydrogen chloride dissolve 65 g of zinc

If

x g of hydrogen chloride dissolve 5 g of zinc

then, from the directly proportional relationship,

$$\frac{x}{73} = \frac{5}{65}$$

$$\therefore \quad x = \frac{5}{65} \times 73 = 5.6 \text{ g}$$

Example

A copper sulphate hydrate has the formula $CuSO_4.yH_2O$, where y is an unknown number. On heating 10.0 g of the hydrate, the mass of anhydrous copper sulphate left was 6.4 g. Calculate y. (Relative atomic mass: $Cu = 64$, $S = 32$, $O = 16$, $H = 1$.)

$CuSO_4$ (anhydrous) has a relative molecular mass $= 64 + 32 + (4 \times 16) = 160$

and yH_2O has a relative molecular mass $= y(2 + 16) = 18y$

Now mass of water H_2O in hydrate $= 10.0 - 6.4 = 3.6$ g

and mass of $CuSO_4$ (anhydrous) $= 6.4$ g

So, by proportion of water H_2O to anhydrous $CuSO_4$, we have

$$\frac{18y}{160} = \frac{3.6}{6.4}$$

$$y = \frac{160 \times 3.6}{18 \times 6.4} = \frac{160 \times 36}{18 \times 64} = 5$$

Volumes of gases

As stated on p. 141, 1 mole of any gas has a volume at s.t.p. of about 22.4 dm^3 (or 22.4 l), which is 22 400 cm^3 (Figure 12.2).

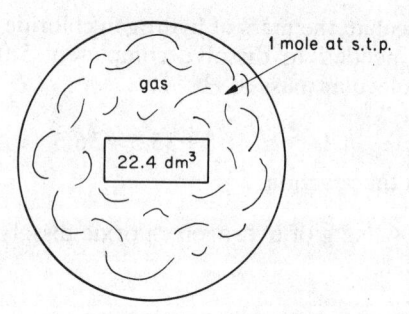

Figure 12.2.

Consider the reaction between zinc and dilute hydrochloric acid which yields hydrogen gas:

$$Zn + 2HCl \rightarrow ZnCl_2 + H_2$$

Thus 1 mole of zinc produces 1 mole of hydrogen gas. Since 1 mole of zinc has a mass of 65 g, and 1 mole of hydrogen at s.t.p. has a volume of 22.4 dm^3, then

65 g of zinc produces 22.4 dm^3 or 22 400 cm^3 of hydrogen at s.t.p.

Suppose

1.3 g of zinc produces x dm^3 of hydrogen at s.t.p.

Then, by proportion,

$$\frac{x}{22.4} = \frac{1.3}{65}$$

$$\therefore \quad x = \frac{1.3}{65} \times 22.4 \text{ dm}^3$$

$$= \frac{1.3}{65} \times 22\,400 \text{ cm}^3 = \frac{13}{650} \times 22\,400 \text{ cm}^3$$

$$= 448 \text{ cm}^3$$

By proportion, we can also calculate the mass of zinc needed to produce 50 cm^3 of hydrogen. If y gram of zinc is required, then

$$\frac{y}{65} = \frac{50}{22\,400}$$

$$\therefore \quad y = \frac{50}{22\,400} \times 65 = 0.15 \text{ g (approx.)}$$

Example

When steam, or vaporized water H_2O, is passed over heated iron filings (Fe), hydrogen gas is formed. The reaction is:

$$3Fe + 4H_2O \rightarrow Fe_3O_4 + 4H_2$$

Calculate the volume of hydrogen produced at s.t.p. if 2 g of iron is used. (Relative atomic mass: Fe = 56, O = 16, H = 1.)

From the reaction,

3 moles of iron produce 4 moles of hydrogen

Now mass of 1 mole of iron = 56 g and 1 mole of hydrogen gas has a volume of 22.4 dm^3 s.t.p. So

3 × 56 or 168 g iron produces 4 × 22.4 or 89.6 dm^3 hydrogen at s.t.p.

Suppose

2 g iron produces x dm^3 hydrogen at s.t.p.

Then, by proportion,

$$\frac{x}{89.6} = \frac{2}{168}$$

$$\therefore \quad x = \frac{2}{168} \times 89.6 = 1.07 \text{ dm}^3 = 1070 \text{ cm}^3$$

EXERCISE 40

1. In the reaction between hydrogen gas and chlorine gas, hydrogen chloride gas is formed:

$$H_2 + Cl_2 \rightarrow 2HCl$$

    Calculate (i) the mass of hydrogen required to produce 3.65 g of hydrogen chloride, (ii) the mass of chlorine which would combine with 0.50 g of hydrogen gas to form hydrogen chloride. (Relative atomic mass: $Cl = 35.5$, $H = 1$.)

2. In the reaction in Question 1, find (i) the volume of hydrogen chloride gas at s.t.p. produced by 20 cm^3 of hydrogen gas at s.t.p., (ii) the volume of hydrogen chloride gas at s.t.p. produced by 0.02 g of hydrogen.

3. In a reaction between calcium (Ca) and water (H_2O), 1.0 g of calcium reacts with 0.9 g of water. Calculate the number of moles of water which react with 1 mole of calcium. (Relative atomic mass: $Ca = 40$, $O = 16$, $H = 1$.)

4. In the reaction below between copper(II) oxide and excess of hydrogen gas, the oxide is reduced to copper and water is formed:

$$CuO + H_2 \rightarrow Cu + H_2O$$

    Calculate (i) the mass of hydrogen which combines with 7.95 g of copper oxide, (ii) the mass of copper then produced, (iii) the volume of hydrogen at s.t.p. which would produce 6.35 g of copper. (Relative atomic mass: $Cu = 63.5$, $O = 16$, $H = 1$.)

5. Crystals of iron(II) sulphate have a formula $FeSO_4.xH_2O$, where x is a number.
    On heating 13.9 g of the crystals, 7.6 g of the anhydrous iron sulphate is left. Calculate x. (Relative atomic mass: $Fe = 56$, $S = 32$, $O = 16$, $H = 1$.)

6. When magnesium (Mg) burns in air, magnesium oxide is formed:

$$2Mg + O_2 \rightarrow 2MgO$$

    Find (i) the mass of oxygen which combines with 1.2 g of magnesium, (ii) the volume of oxygen at s.t.p. which would produce 2 g of magnesium oxide, (iii) the mass of magnesium which would produce 8 g of magnesium oxide. (Relative atomic mass: $Mg = 24$, $O = 16$.)

ELECTROLYSIS

Facts about the mole

1. 1 mole is the amount of substance which has the same number of particles (for example, atoms or molecules or ions or electrons) as the number of atoms in 12 g of carbon 12. This number is about 6×10^{23} and is known as the *Avogadro constant*.
2. 1 mole of any element or compound has the same number of molecules, about 6×10^{23}.
3. 1 mole of *any* gas has the same volume, about 22.4 dm^3 (22.4 l), at s.t.p.

Relative atomic mass; molar mass

Monatomic element Silver, symbol Ag, is a typical monatomic element; its molecules consist of 1 atom. Silver has a relative atomic mass of 108; so the mass of 1 mole of silver, or its molar mass, is 108 g.

Diatomic element Hydrogen, symbol H, is a diatomic element; its molecules consist of 2 atoms. So the chemical formula for the molecule is H_2. The relative atomic mass of hydrogen is 1; so its relative molecular mass is 2 and hence the molar mass of hydrogen is 2 g.

The table below shows some typical elements together with their relative atomic mass, atomicity (number of atoms in molecule), and molar mass.

Element	Rel. atomic mass	Atomicity	Molecular formula	Molar mass (g)
Silver	108	1	Ag	108
Copper(II)	63.5	1	Cu	63.5
Aluminium	27	1	Al	27
Hydrogen	1	2	H_2	2
Oxygen	16	2	O_2	32
Chlorine	35.5	2	Cl_2	71

Thus 2 g of hydrogen, 32 g of oxygen and 71 g of chlorine all occupy a volume of 22.4 dm^3 at s.t.p.

Electrolysis

1. 1 F (faraday) is the charge required to liberate in electrolysis 1 mole of a monovalent element such as silver.

2. In electrolysis, the charge is carried through the electrolytic solution *by the ions* in solution.

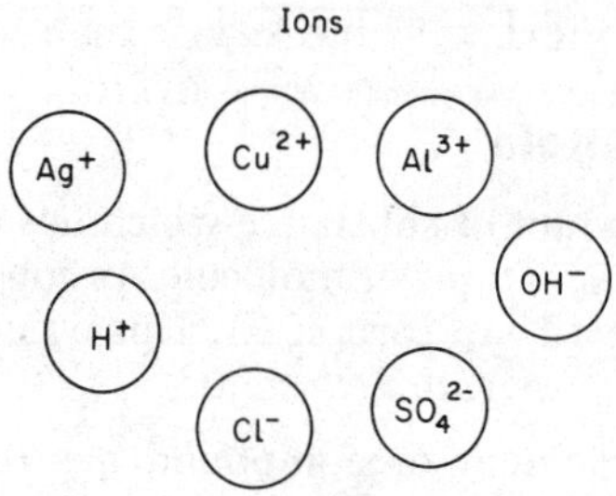

Ions

Figure 12.3.

The table below shows some typical ions and the charge they carry. Ag^+ indicates that a silver ion has a positive charge $+1e$, where e is the electronic charge; Cu^{2+} indicates that a copper(II) ion has a positive charge $+2e$; Cl^- indicates a chlorine ion with a negative charge $-1e$. Generally, the ion of a monovalent atom carries a charge $1e$, the ion of a divalent atom carries a charge $2e$, and so on (Figure 12.3).

	Atom	Ion
Metals	Silver Copper(II) Aluminium	Ag^+ Cu^{2+} Al^{3+}
Gases	Hydrogen Chlorine	H^+ Cl^-

3. Metals Since the charge is carried by the ions themselves, it follows from the table that, in electrolysis,

1 F liberates 1 mole of silver, 108 g

or $\frac{1}{2}$ mole of copper(II), 31.75 g

or $\frac{1}{3}$ mole of aluminium, 9 g

4. Gases (a) 1 F liberates 1 mole of hydrogen (H) atoms. They form $\frac{1}{2}$ mole of hydrogen gas molecules (H_2), since 2 atoms bond to form 1 molecule. So

mass of hydrogen liberated by $1 F = \frac{1}{2} \times 2$ g $= 1$ g

The *volume* of gas liberated by $1 F = \frac{1}{2}$ mole $= \frac{1}{2} \times 22.4 = 11.2$ dm^3 at s.t.p.

(b) 1 F also liberates $\frac{1}{2}$ mole of chlorine gas, or 11.2 dm^3 of the gas at s.t.p.

(c) Since

$$2H_2O \to 2H_2 + O_2$$

we see that 1 mole of oxygen is liberated when 4 moles of hydrogen atoms (H) are liberated. But 4 moles of hydrogen atoms are liberated by a charge of 4 F. Hence

4 F liberates 1 mole of oxygen gas or 22.4 dm^3 at s.t.p.

So

1 F liberates $\frac{1}{4}$ mole of oxygen or 5.6 dm^3 at s.t.p.

EXERCISE 41

(Where necessary, assume that 1 F = 96 500 C and that 1 mole of any gas occupies a volume at s.t.p. of 22.4 dm^3.)

1. (i) Write down the symbols for the *ions* of hydrogen, silver, copper(II), chlorine, aluminium. (ii) What charge does each ion carry? (iii) How many moles of each element is liberated in electrolysis by 1 F?

2. 1 mole of a monovalent element such as silver is liberated in electrolysis by a charge of 1 F. (i) Why is $\frac{1}{2}$ mole of a divalent element X, and $\frac{1}{3}$ mole of a trivalent element Y, liberated by 1 F? (ii) Name a metal for X and for Y. (iii) What volume of hydrogen at s.t.p. is produced by 1 F in electrolysis?

3. Calculate (i) the mass of silver, (ii) the mass of hydrogen, (iii) the volume of hydrogen at s.t.p. liberated in electrolysis by the passage of 9650 C. (Relative atomic mass: silver = 108, hydrogen = 1.)

4. What is the charge in faraday, F, required to liberate in electrolysis (i) 13.5 g of aluminium, (ii) 0.1 g of hydrogen, (iii) 5.6 dm^3 of oxygen at s.t.p.? (Relative atomic mass: aluminium = 27, oxygen = 16, hydrogen = 1.)

5. When a charge of 4825 C is passed through a solution of copper(II) sulphate solution with copper electrodes, 1.59 g of copper is liberated. Calculate the mass of 1 mole of copper.
   What mass of silver would be liberated by a charge of 4825 C in electrolysis? (Relative atomic mass: silver = 108.)

6. In the electrolysis of dilute sulphuric acid solution, 112 cm^3 at s.t.p. of hydrogen is collected. Calculate the charge in F passing through the solution.

7. A steady current of 0.5 A flows for 30 min through a solution of copper(II) sulphate solution with copper electrodes. (i) What quantity of charge passes through the solution? (Quantity = current in amperes × time in *seconds*, where quantity is in coulomb C.) (ii) Calculate the mass of copper liberated. (Relative atomic mass of copper = 63.5.)

TRIGONOMETRY

13 Basic Trigonometry

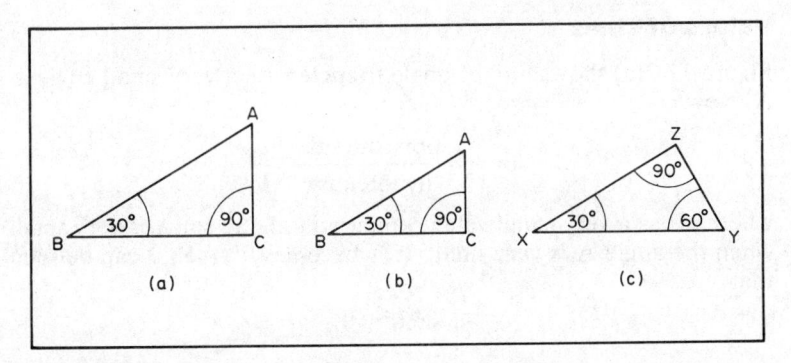

Figure 13.1.

SINE AND APPLICATIONS IN OPTICS

Sine of angle

Figure 13.1(a) shows a *right-angle* triangle ABC with an angle of 30° at B. The hypotenuse or longest side faces the right-angle corner C and is AB.

In any right-angle triangle, the *sine* of an angle in it is defined as the ratio

$$\frac{\text{side opposite angle}}{\text{hypotenuse}}$$

We write 'sin' in short for 'sine'. So in Figure 13.1(a)

$$\sin 30° = \frac{AC}{AB}$$

Suppose the triangle ABC shrinks to a smaller size but still has the same angles, as shown in Figure 13.1(b). The ratio AC/AB for sin 30° will then have the same value as before. So the sine of an angle is a *constant*—it does not depend on the size of the triangle containing the angle.

148 BASIC MATHEMATICS FOR SCIENCE

Figure 13.1(c) shows a right-angle triangle XYZ containing an angle of 60° at Y. The hypotenuse, the side opposite the 90° corner Z, is XY. Then

$$\sin 60° = \frac{\text{opposite side}}{\text{hypotenuse}} = \frac{XZ}{XY}$$

and

$$\sin 30° = \frac{\text{opposite side}}{\text{hypotenuse}} = \frac{YZ}{XY}$$

Values of sines

Figure 13.2(a) shows a right-angle triangle with a very small angle at A. Now

$$\sin A = \frac{\text{opposite side}}{\text{hypotenuse}} = \frac{a}{h}$$

where a has a very small value compared to h. So sin A is very small when the angle A is very small. If A becomes 0°, then it can be seen that

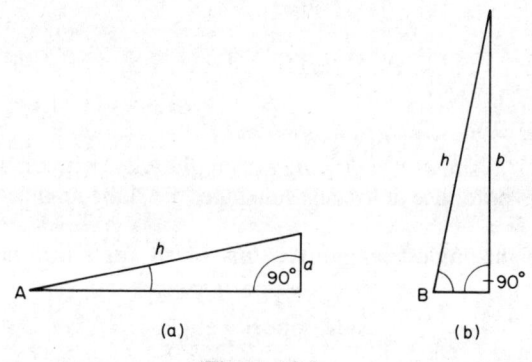

$$\sin 0° = 0$$

Figure 13.2.

In Figure 13.2(b) however, angle B is very large and nearly 90°. Then

$$\sin B = \frac{b}{h}$$

and since b is nearly equal to h in length, it follows that the sine is nearly 1 in value. If B becomes 90°, then it can be seen that

$$\sin 90° = 1$$

BASIC TRIGONOMETRY

All the sine values of angles between 0° and 90° are listed to four or more figures in trigonometrical tables. 'Log sin' tables give their logarithms directly for use in calculations. The sine values of the special angles 30°, 45° and 60° can easily be calculated from right-angle triangles containing these angles, as we now show.

Sines of 30°, 45°, and 60°

In Figure 13.3(a), ABC is an equilateral triangle whose sides are each 2 units long. The perpendicular AN from A to BC bisects BC, so that BN is 1 unit long. From the right-angle triangle ABN the side $AN = \sqrt{2^2 - 1^2} = \sqrt{3}$ by Pythagoras. So

$$\sin 60° = \frac{\text{opposite}}{\text{hypotenuse}} = \frac{AN}{AB} = \frac{\sqrt{3}}{2} = 0.866 \quad (3 \text{ s.f.})$$

and

$$\sin 30° = \frac{\text{opposite}}{\text{hypotenuse}} = \frac{BN}{AB} = \frac{1}{2} = 0.5$$

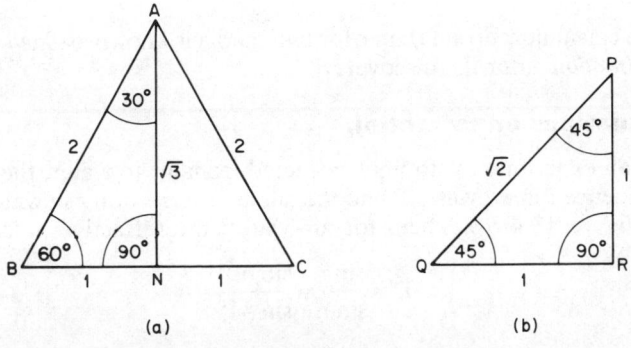

Figure 13.3.

In Figure 13.3(b), the right-angle triangle PQR has an angle of 45°, since PR = QR. If PR = 1 unit, then, from Pythagoras,

$$PQ = \sqrt{1^2 + 1^2} \equiv \sqrt{2} = \text{hypotenuse}$$

So

$$\sin 45° = \frac{\text{opposite}}{\text{hypotenuse}} = \frac{1}{\sqrt{2}} = 0.707$$

Sine law of refraction

When light is *refracted* at a plane boundary from one medium such as air to another medium such as glass (Figure 13.4(a)), the angles of incidence i and refraction r are related to each other by

$$\frac{\sin i}{\sin r} = n$$

where n is a constant. For air–glass, n is the 'refractive index' from air to glass, and this has the value about 1.5 for crown glass (Figure 13.4(a)).

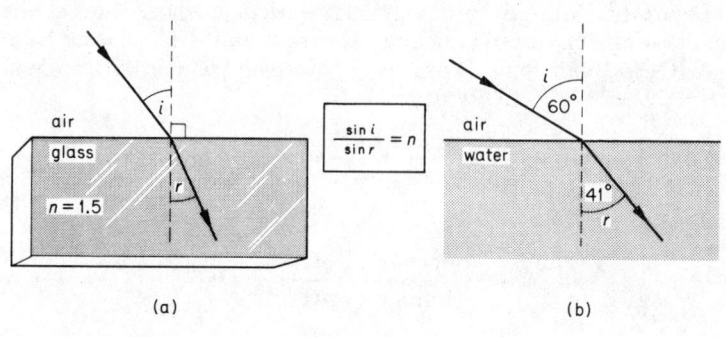

Figure 13.4.

The constant ratio $\sin i/\sin r$ for two media is known as *Snell's Law of Refraction*, after the discoverer.

Calculations on refraction

1. In an experiment with light refracted from air to water, the angle of incidence i in air was 60° and the angle of refraction r in water was 41° (Figure 13.4(b)). Then, for air–water, the refractive index n is given by

$$n = \frac{\sin i}{\sin r} = \frac{\sin 60°}{\sin 41°}$$

$$= \frac{0.866}{0.656} = 1.32$$

2. Suppose the angle of refraction r in glass from air is 30°, the refractive index of air–glass is 1.5, and we need to find the angle of incidence i in the air (Figure 13.5(a)). Then, from Snell's Law,

$$\frac{\sin i}{\sin r} = n$$

so

$$\frac{\sin i}{\sin 30°} = 1.5$$

∴ $\sin i = 1.5 \times \sin 30° = 1.5 \times 0.5 = 0.75$

From tables,
$$i = 49°$$

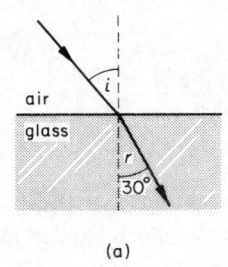

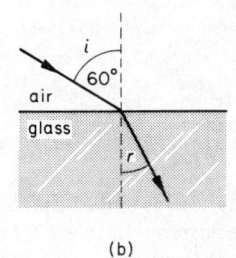

Figure 13.5.

3. We can also find the angle of refraction r from $\sin i / \sin r = n$. For example, suppose the angle of incidence i in air is 60° and the angle of refraction r in glass of $n = 1.5$ is required (Figure 13.5(b)). Then

$$\frac{\sin 60°}{\sin r} = 1.5$$

$$\therefore \frac{0.866}{\sin r} = 1.5$$

$$\therefore 0.866 = 1.5 \times \sin r$$

$$\therefore \sin r = \frac{0.866}{1.5} = 0.577$$

From tables,
$$r = 35°$$

Critical angle

When light travels from glass to air it is always refracted *away* from the normal at the point of incidence (Figure 13.6(a)). This is also the case when light travels from water to air or, in general, from one medium to another which is optically *less dense*.

For one particular angle of incidence C in glass, the refracted ray will travel along the boundary OB between the glass and the air (Figure 13.6(b)). When the angle of incidence is now increased, however slightly, *no refracted ray appears. All the light is then reflected* at O into the glass. In Figure 13.6(a), however, when the angle of incidence is less than C, only a small amount of light is reflected back into the glass; most of the light is refracted.

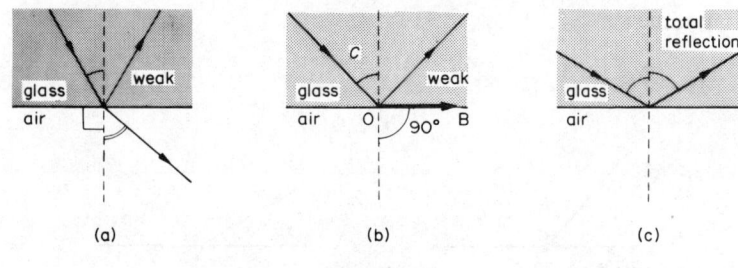

Figure 13.6.

The angle C is called the *critical angle* for glass–air. For an angle of incidence *greater* than C, *total internal reflection* is said to take place because all the light is then reflected back inside the glass.

Calculation of critical angle

We can calculate the critical angle for air–glass by using Snell's Law. Suppose the glass has a refractive index n of 1.5 and C is the critical angle. We can see from Figure 13.6(b) that a ray travelling along BO *in air* would be refracted into the glass at an angle of refraction C. In this case the angle of incidence in the air is 90°. So

$$\frac{\sin 90° \ (i)}{\sin C \ (r)} = n = 1.5$$

Now $\sin 90° = 1$ (p. 148). So

$$\frac{1}{\sin r} = 1.5$$

$$\therefore \quad 1 = 1.5 \times \sin r$$

$$\therefore \quad \sin r = \frac{1}{1.5} = \frac{10}{15} = 0.6667$$

From tables,

$$r = 42°$$

This result shows that if the angle of incidence of light in glass is greater than 42°, then total internal reflection occurs. The effect is used in total reflecting prisms. For water–air, where $n = 1.33$, similar calculation shows that the critical angle is now about 49°.

EXERCISE 42

1. Find sin A and sin B in Figure 13.7(a).

2. In Figure 13.7(b), what is sin A and sin B?

BASIC TRIGONOMETRY 153

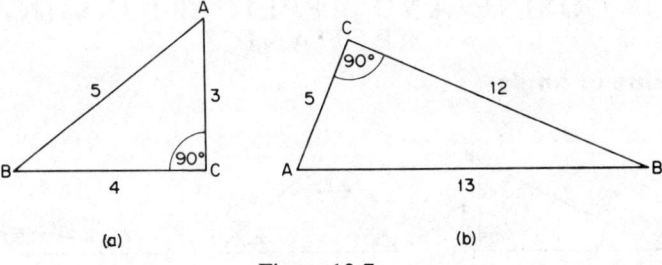

Figure 13.7.

3. Figure 13.8(a) shows one end of a rod 2.00 m long resting on a box of height h. If the rod is at an angle of 20° to the horizontal, find h.
   The rod is now moved so that it makes an angle of 30° with the horizontal. What is the new length of rod between the ground and the top of the box?

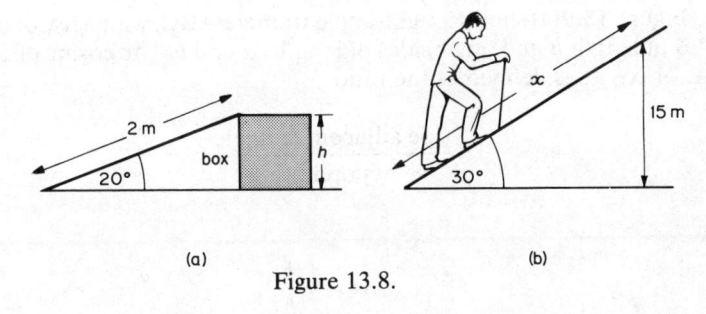

Figure 13.8.

4. A road slopes at 30° to the horizontal (Figure 13.8(b)). If a walker is 15 m higher after travelling a distance x along the road, calculate x.
   If the walker moves on a further distance of 20 m, by what additional height is he raised?

5. A ray of light is incident in air at 45° on an air–glass boundary. The refractive index of the glass is 1.5. Calculate the angle of refraction and draw a ray sketch.

6. A ray of light is incident at an angle i on an air–water boundary, and the angle of refraction in the water is 30°. Find i if the refractive index of water is $\frac{4}{3}$.

7. The refractive index of flint glass is 1.7. Calculate the critical angle for this glass–air boundary. Draw a ray sketch in illustration.

8. The critical angle for a liquid–air boundary is 45°. Find the refractive index of the liquid.

COSINE AND APPLICATIONS IN MECHANICS

Cosine of angle

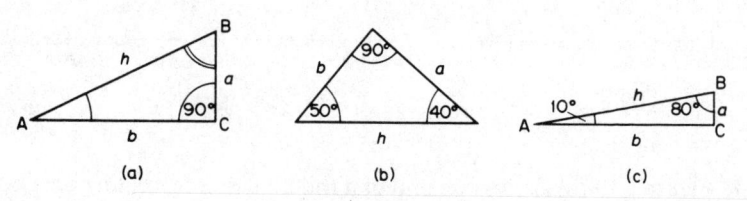

Figure 13.9.

1. We now consider another trigonometrical ratio called the *cosine*, or 'cos' for short.

Figure 13.9(a) shows a right-angle triangle ABC, with a hypotenuse AB of length h and other sides of lengths a and b. The cosine of angle A, or cos A, is defined as the ratio

$$\frac{\text{side adjacent to angle}}{\text{hypotenuse}}$$

so

$$\cos A = \frac{b}{h}$$

and

$$\cos B = \frac{a}{h}$$

In the right-angle triangle shown in Figure 13.9(b)

$$\cos 50° = \frac{b}{h}$$

and

$$\cos 40° = \frac{a}{h}$$

2. Figure 13.9(c) shows a triangle with a small angle A of only about 10° and a large angle B of about 80°. Now

$$\cos A = \frac{b}{h}$$

BASIC TRIGONOMETRY

and *b* is nearly as long as *h*. So cos A has a value *b/h* such as 0.98, or nearly 1. In the limit, when angle A is 0°, cos 0° = 1.

Conversely,

$$\cos B = \frac{a}{h}$$

and *a* is extremely small compared with *h*. So cos B has a value such as 0.17, which is very small. In the limit, when angle B is 90°, cos 90° = 0. Cosines of all the angles between 0° and 90° can be found to at least four decimal places from trigonometrical tables.

Cosine values for 30°, 45°, and 60°

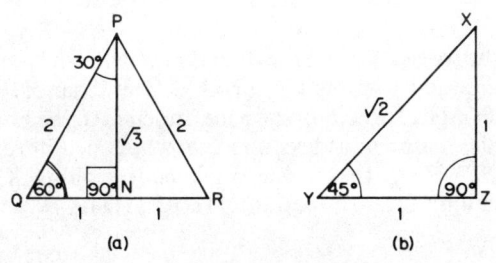

Figure 13.10.

Figure 13.10(a) shows an equilateral triangle PQR of side 2 units. PN is a perpendicular from P to QR so that angle N = 90° and QN = NR = 1 unit. By Pythagoras for triangle PQN, PN = $\sqrt{3}$ (p. 149).

From the right-angle triangle PQN it can be seen that

$$\cos 30° = \frac{PN}{PQ} = \frac{\sqrt{3}}{2} = 0.866$$

and

$$\cos 60° = \frac{QN}{PQ} = \frac{1}{2} = 0.5$$

Figure 13.10(b) shows a right-angle triangle XYZ with sides XZ and YZ equal to 1 unit. So angle Y = angle X = 45°. Also, from Pythagoras, XY = $\sqrt{2}$. So

$$\cos 45° = \frac{1}{\sqrt{2}} = 0.707$$

Components of forces

Forces are examples of a group of quantities called *vectors*. Vectors can be represented in magnitude (size) and direction by a straight line drawn to scale. For example, Figure 13.11(a) shows a line OB, 10 cm long to a scale of 1 cm = 10 N; it represents a force F of 100 N in a direction at 30° to a direction OC.

(a)

(b)

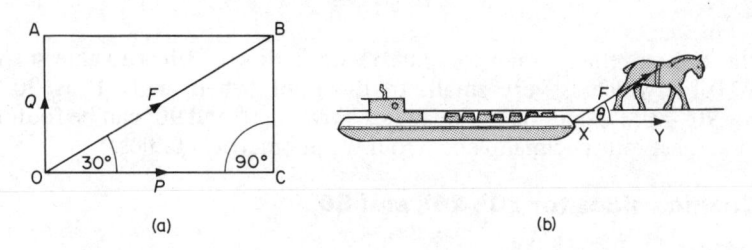

Figure 13.11.

Although the force F acts in the direction OB, it can have an effect in a different direction such as OC or OA. One example is shown in Figure 13.11(b). The force F in the rope attached to the horse pulls the barge along the water in a direction XY which is different from the direction of F. We say that F has a *component* along XY in Figure 13.11(b) and along OC or OA in Figure 13.11(a).

Magnitude of component

We can find the magnitude of the component of F in Figure 13.11(a) by drawing a *rectangle* OABC which has OB (F) as its diagonal. By vector theory, the forces P and Q, represented by the sides OC and OA, together equal F. In other words, F is the 'resultant' of P and Q when these forces are added together. Now the force Q has no effect in the direction OC, because OC is perpendicular to the direction of Q. So P is the effective part or component of F in the direction OC.

From the right-angle triangle OBC in Figure 13.11(a), we see that

$$\frac{OC}{OB} = \cos 30°, \quad \text{so} \quad \frac{P}{F} = \cos 30°$$

$$\therefore \quad P = F \cos 30°$$

So if $F = 100$ N, its component in the direction OC is

$$P = 100 \cos 30°$$
$$= 100 \times 0.866 = 86.6 \text{ N}$$

We can now give a general rule for the magnitude of the component of a force F. In a direction making an angle θ with the direction of F,

$$\text{component} = F \cos \theta$$

BASIC TRIGONOMETRY

Using this rule, we see that the component Q in the direction 60° to F is given by

$$Q = F \cos 60° = 100 \times 0.5 = 50 \text{ N}$$

In a direction perpendicular to F, $\theta = 90°$. So

$$\text{component} = F \cos 90° = F \times 0 = 0$$

Forces in equilibrium

1. Consider a block O on a smooth plane inclined at 30° to the horizontal (Figure 13.12(a)). It is held from slipping down by a rope attached to it, which exerts an upward force F along the plane as shown.

The weight W of the block acts vertically downwards. *The component of W down the plane* tries to pull the block down the plane. Since the block is stationary, it follows that $F =$ component of W at 60° to its direction. So

$$F = W \cos 60°$$

Thus if W is 200 N,

$$F = 200 \cos 60° = 200 \times 0.5 = 100 \text{ N}$$

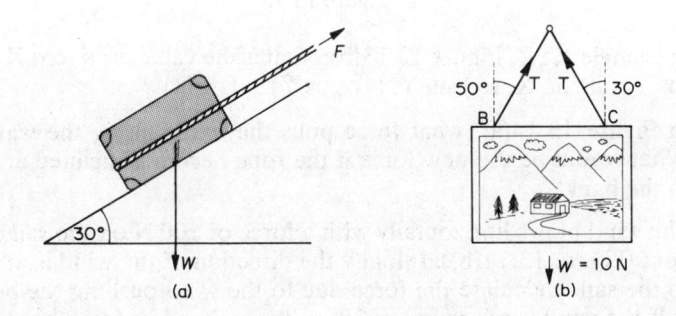

Figure 13.12.

2. Consider a picture of weight W suspended by a string attached to its corners at B and C as shown in Figure 13.12(b). Suppose W is 10.0 N. Then the weight of 10 N, which acts vertically downwards, is supported by the *vertical components* of the force or tension T in each string which act at B and C as shown.

Now the vertical component of T at B or C is $T\cos 30°$, if the string is inclined at 30° to the vertical at B or C. So

$$T\cos 30° + T\cos 30° = 10$$
$$\therefore\ 2T\cos 30° = 10$$
$$\therefore\ T = \frac{10}{2\cos 30°}$$
$$= \frac{10}{2 \times 0.866}$$
$$= 5.8\text{ N}$$

EXERCISE 43

1. In triangle ABC, Figure 13.13(a), write down the value of (i) cos A, (ii) cos B.

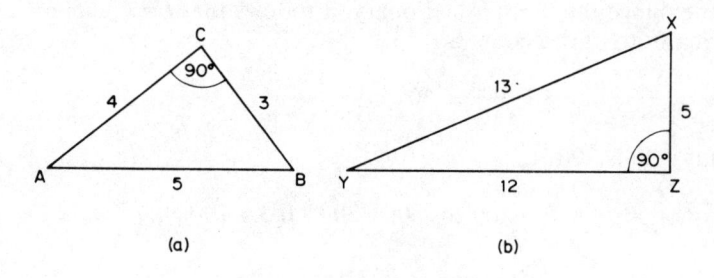

Figure 13.13.

2. In triangle XYZ, Figure 13.13(b), what is the value of: (i) cos X, (ii) cos Y, (iii) sin X, (iv) sin Y, (v) $(\cos Y)^2 + (\sin Y)^2$?

3. In Figure 13.14(a), what force pulls the barge along the water? What would be the new force if the rope becomes inclined at 30° to the bank?

4. The wind blows horizontally with a force of 100 N on the sail of a boat (Figure 13.14(b)). Initially the direction of the wind is at 60° to the sail. Calculate the force due to the wind pushing the boat.
   If the wind veers round at 30° to the sail and its force becomes 200 N, calculate the new force due to the wind pushing the boat.

5. A load W of 200 N is supported by two strings inclined at 60° to the horizontal (Figure 13.14(c)). If the force in each string is F, calculate F.

6. A block X is placed on a smooth inclined plane inclined at 60° to the horizontal. The weight of X is 100 N and X is just kept in

BASIC TRIGONOMETRY 159

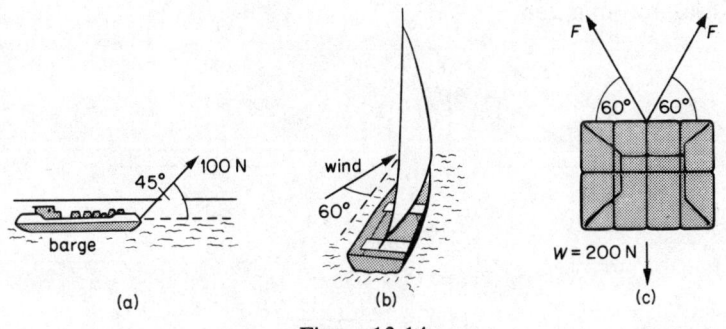

Figure 13.14.

position by a horizontal force F as shown (Figure 13.15(a)). Find (i) the component of F up the plane, (ii) the component of W down the plane, (iii) the magnitude of F.

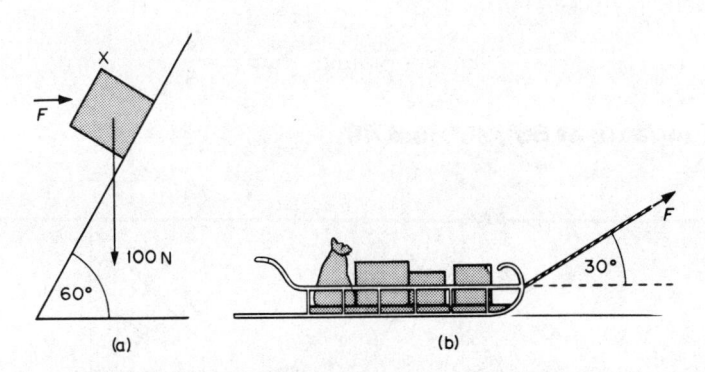

Figure 13.15.

7. A rope attached to a sledge is inclined at 30° to the horizontal (Figure 13.15(b)). The force F in the rope is 80 N. Calculate the force pulling the sledge along the ground. What is the vertical component of F?

If the force pulling the sledge along the ground changes to 50 N while the force in the rope remains 80 N, what is now the angle of inclination of the rope to the horizontal?

TANGENT AND APPLICATIONS IN VECTORS

Tangent of angle

In addition to sine and cosine, the *tangent* of an angle is used. We write this 'tan' for short. In the right-angle triangle ABC, Figure

13.6(a), by definition,

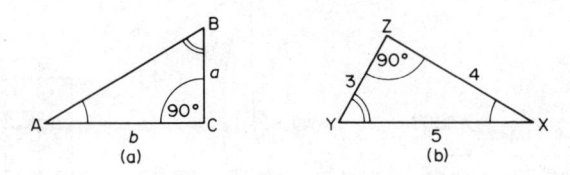

Figure 13.16.

and

$$\tan A = \frac{\text{opposite}}{\text{adjacent}} = \frac{a}{b}$$

$$\tan B = \frac{\text{opposite}}{\text{adjacent}} = \frac{b}{a}$$

In Figure 13.16(b),

$$\tan X = \frac{3}{4} \quad \text{and} \quad \tan Y = \frac{4}{3}$$

Tangents of 60°, 30°, and 45°

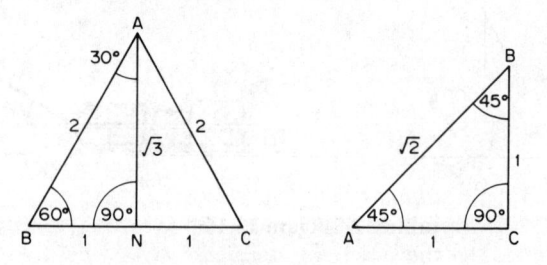

Figure 13.17.

Figure 13.17(a) shows an equilateral triangle of side 2 units and a perpendicular AN to BC so that BN = 1 unit. Then, as shown on p. 149, AN = $\sqrt{3}$. From triangle ABN, we have

$$\tan 60° = \frac{AN}{BN} = \frac{\sqrt{3}}{1} = \sqrt{3} = 1.732$$

and

$$\tan 30° = \frac{BN}{AN} = \frac{1}{\sqrt{3}} = 0.577$$

Figure 13.17(b) shows an isosceles right-angle triangle ABC, with BC = AC = 1 unit and angle A = 45° = angle B. As shown on p. 149,

$AB = \sqrt{2}$ units. From triangle ABC,

$$\tan 45° = \frac{AC}{BC} = \frac{1}{1} = 1$$

Perpendicular vectors; resultant of velocities

Consider a boy or girl swimming across a stream which flows at 2 m/s due east (Figure 13.18(a)).

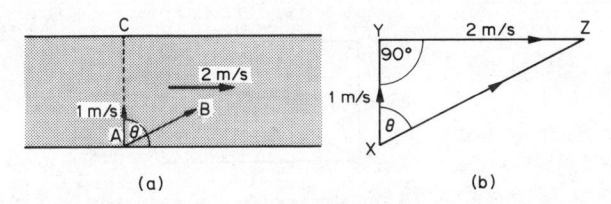

Figure 13.18.

In still water the velocity of the swimmer may be 1 m/s and he or she would swim perpendicular to the bank to reach the opposite side. But the current carries the swimmer downstream at 2 m/s. So instead of moving in the direction AC, the swimmer actually moves in the direction AB at an angle θ to AC.

A velocity is a *vector*, that is, it has magnitude (size) and direction. To find the angle θ, we draw a vector XY to represent the velocity of 1 m/s of the swimmer and then add the vector YZ perpendicular to XY to represent the velocity of 2 m/s of the stream (Figure 13.18(b)). The *resultant* of the two velocities is the vector XZ. So the direction of the swimmer is the direction of XZ. XZ also represents the actual velocity of the swimmer in the stream.

Suppose XZ makes an angle θ with XY. (This angle is also shown in Figure 13.18(a).) Then, since angle Y is 90°, it follows that

$$\tan \theta = \frac{YZ}{XY} = \frac{2}{1} = 2$$

From tables:

$$\theta = 63°$$

We can also find the resultant velocity v of the swimmer, which is XZ. From triangle XYZ,

$$\frac{YZ}{XZ} = \sin \theta$$

So
$$\frac{2}{v} = \sin 63° = 0.89$$
$$\therefore \quad 2 = v \times 0.89$$
$$\therefore \quad v = \frac{2}{0.89} = 2.2 \text{ m/s}$$

Forces and their resultant

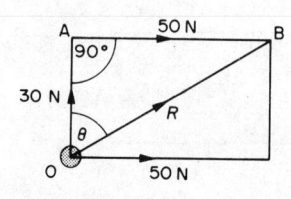

Figure 13.19.

Consider two perpendicular forces of 30 N and 50 N pulling an object at O (Figure 13.19). If OA is drawn to represent the 30 N force and AB to represent the 50 N force, then OB is the resultant force. The object at O will move in the direction OB with a force equal in magnitude to OB.

Since angle A is 90°,
$$\tan \theta = \frac{AB}{OA} = \frac{50}{30} = 1.667$$

From tables:
$$\theta = 59°$$

So the object would move in a direction at 59° to OA.

The magnitude R of the resultant force can be found from triangle OAB. We have
$$\frac{AB}{OB} = \sin \theta$$

So
$$\frac{50}{R} = \sin 59° = 0.86$$
$$\therefore \quad 50 = R \times 0.86$$
$$\therefore \quad R = \frac{50}{0.86} = 58 \text{ N}$$

BASIC TRIGONOMETRY   163

Similar calculations apply not only to velocities and forces but to all vectors such as acceleration, momentum and electric and magnetic field strengths. In contrast, quantities such as mass, density, work and energy have only magnitude or numerical values; they have no direction. These quantities are called *scalars*.

EXERCISE 44

1. In Figure 13.20(a), find (i) tan A, (ii) tan B.

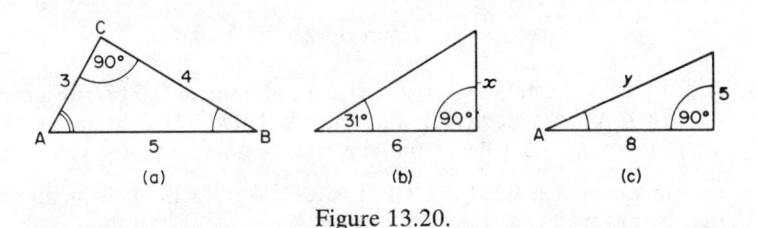

Figure 13.20.

2. In Figure 13.20(b), find the length x.

3. In Figure 13.20(c), calculate (i) angle A, (ii) length y.

4. A ship travels 2.0 km due north from X to Y, and then travels 1.5 km due east from Y to Z (Figure 13.21(a)). Find (i) the angle XZ makes with the initial direction XY, (ii) the distance XZ.

5. The forces in two perpendicular ropes supporting a weight W are 20 N and 30 N respectively (Figure 13.21(b)). Calculate (i) the angle between their resultant R and the 20 N force, (ii) the resultant R.
   What is the magnitude of W?

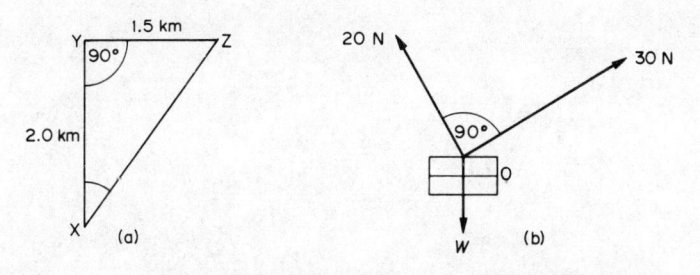

Figure 13.21.

6. An aeroplane travelling at 800 km/h encounters a 100 km/h wind at right angles to its direction. Find (i) the angle which the aeroplane is off course, (ii) the resultant velocity of the aeroplane.

7. Two perpendicular forces P and Q have a resultant R which acts at 50° to P (Figure 13.22(a)). If Q is 60.0 N, find (i) P, (ii) R.

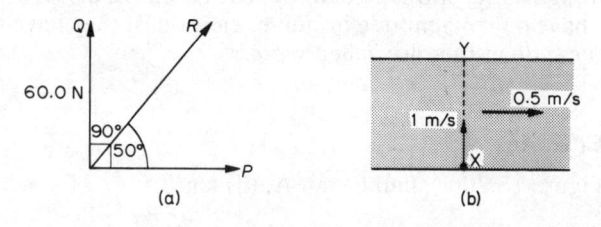

Figure 13.22.

8. A boy swims perpendicular to a stream moving at 0.5 m/s (Figure 13.22(b)). If his velocity in still water is 1.0 m/s, calculate (i) the angle θ to XY that he actually swims, (ii) his resultant velocity.

9. In Question 8, in what direction to XY should the boy swim so that he always heads in the direction XY?

14 Revision Papers

PAPER 1

1. Find the value of (i) $3\frac{1}{8} - 1\frac{1}{2}$, (ii) $3\frac{1}{3} - 1\frac{4}{5} + 2\frac{1}{6}$.

2. Calculate (i) 2.6×0.05, (ii) $0.08 \div 0.2$.

3. Solve the equations (i) $5x - 4 = 2x + 5$, (ii) $\frac{3x}{4} + \frac{1}{3} = \frac{x}{6} + 1$.

4. If $P = I^2 R$, (i) calculate P if $I = 2$, $R = 10$, (ii) calculate I if $P = 20$, $R = 5$.

5. The volume of a gas collected at 17°C (290 K) and 750 mmHg pressure is 44.0 cm³. Calculate the volume at s.t.p. (0°C and 760 mmHg).

6. The period T of a simple pendulum is proportional to the square root of its length l. If $T = 2.00$ s when $l = 1.0$ m, find (i) the period when l is 1.5 m, (ii) the length which gives a period of 3.50 s.

7. (i) In triangle XYZ, calculate the lengths YZ and XZ if XY = 5.0 cm (Figure 14.1(a)).

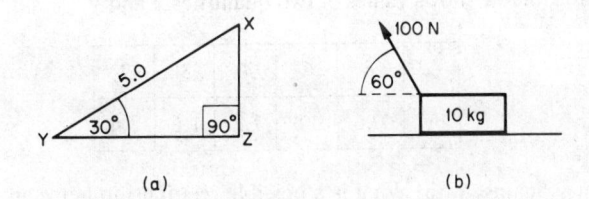

Figure 14.1.

(ii) An object of mass 10 kg, resting on a horizontal smooth plane, is suddenly pulled by a force of 100 N acting at an angle of 60° to the horizontal (Figure 14.1(b)). (a) Find the component of the force in the horizontal direction. (b) If $F = ma$, where F is the force acting on an object of mass m and a is the acceleration produced, calculate the initial acceleration of the object.

8. In an experiment, measurements are made on the p.d. V and current I for a resistance R. The results are shown in the table.

V (V)	0	1.0	1.5	2.0	2.5	3.0
I (A)	0	0.6	0.9	1.1	1.5	1.8

(i) Plot a graph of V against I. (ii) From your graph, what is the relationship between V and I? (iii) Which measurement in the table would you re-check? (iv) If $R = V/I$, calculate the value of R to 2 significant figures.

PAPER 2

1. Calculate (i) $1\frac{5}{6} \times 1\frac{1}{3}$, (ii) $\dfrac{2\frac{1}{4}}{1\frac{1}{2}}$.

2. Calculate to 2 significant figures (i) 0.067×0.7, (ii) $\dfrac{0.64}{1.2}$.

3. Using logs, find the values of (i) $\dfrac{0.635}{0.076}$ (2 s.f.), (ii) $\dfrac{6.84 \times 0.73^2}{0.0386}$ (3 s.f.).

4. Solve the equations (i) $y + 2 - 2(y - 4) = 3y$, (ii) $\dfrac{1}{v} + \dfrac{1}{10} = \dfrac{1}{12}$.

5. What is the value of (i) $2 \times 10^3 \times 5 \times 10^2$, (ii) $(6 \times 10^{-4})/(3 \times 10^{-2})$?

6. (i) If $v = u + at$, find a in terms of v, u and t. (ii) If $p = h\rho g$, calculate h if $p = 1.2 \times 10^5$, $g = 10$, $\rho = 1000$.

7. The frequency f of a wave is inversely-proportional to the wavelength λ. If $f = 6$ Hz when $\lambda = 0.30$ m, calculate (i) the frequency when λ is 0.20 m, (ii) the wavelength when f is 4 Hz.

8. The table below shows values of two quantities x and y.

x	8	6	4	3	2
y	0.6	0.8	1.2	1.6	2.4

(i) Plot y against x. (ii) What is a possible relationship between y and x? (iii) By plotting another graph, verify your answer to (ii).

PAPER 3

1. Calculate (i) $\dfrac{8\frac{4}{5}}{3\frac{1}{7}}$, (ii) $3\frac{1}{3} \div \frac{5}{6} \times 1\frac{1}{4}$.

2. Calculate to 2 significant figures (i) 1.9×0.08, (ii) $\dfrac{1.9}{0.08}$.

REVISION PAPERS

3. By logs, find to 2 significant figures (i) $\sqrt{0.536}$, (ii) $\dfrac{0.038^2 \times 16.34}{0.185}$.

4. Solve the equations (i) $2a + 5(3-2a) = 2(5-3a)$, (ii) $-\dfrac{1}{12} + \dfrac{1}{u} = -\dfrac{1}{16}$.

5. The momentum of an object is mv, where m is the mass and v is the velocity; the kinetic energy of the object is $\tfrac{1}{2}mv^2$.

   A ball of mass 0.2 kg and velocity 10 m/s hits another object and then moves with a smaller velocity of 4 m/s in the same direction as before. Calculate (i) the change in momentum of the ball, (ii) its change in kinetic energy.

6. The current I flowing in a circuit of resistance R is given by $I = \dfrac{E}{R+r}$, where E is the e.m.f. of the battery supply and r is its internal resistance.
   (i) Calculate I if $E = 4$ V, $R = 8\,\Omega$, $r = 4\,\Omega$. (ii) Find r if $E = 12$ V, $R = 6\,\Omega$ and $I = 1.5$ A.

7. The resistance of a given length of metal is inversely-proportional to the square of its diameter. When the diameter is 0.50 mm, the resistance is 8 Ω. (i) What is the resistance when the diameter is 0.25 mm? (ii) Calculate the diameter when the resistance is 1 Ω.

8. (i) In Figure 14.2(a) calculate the length x of AB and the length y of AC.

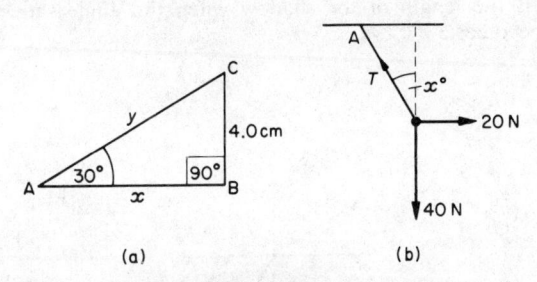

Figure 14.2.

(ii) An object O of weight 40 N is suspended by a string from a point A on the ceiling (Figure 14.2(b)). It is then pulled sideways by a horizontal force of 20 N. Find (a) the resultant of the two forces 20.0 N and 40.0 N, (b) the force T in the string, (c) the direction of T to the vertical, $x°$.

PAPER 4

1. Calculate (i) $2\tfrac{1}{4} - 1\tfrac{3}{8} + 3\tfrac{1}{5}$, (ii) $\dfrac{2\tfrac{1}{4} \times 3\tfrac{1}{3}}{1\tfrac{3}{5}}$.

2. Find the value of (i) 0.075×1.2, (ii) $\dfrac{2.4}{0.25}$, (iii) to 2 s.f. using logs, $\sqrt{\dfrac{9.64}{0.572}}$.

3. (i) Calculate the percentage of iron Fe in the compound Fe_2S_3 if the relative atomic mass = Fe 56, S 32. (ii) The percentage composition by mass of sodium nitrate is Na 27%, N 16.5% and O 56.5%. If the relative atomic mass = Na 23, N 14, O 16, find the empirical formula of sodium nitrate.

4. Solve the equations

(i) $3(R+2) = 10 - R$;

(ii) $\dfrac{1}{12} + \dfrac{1}{8} = \dfrac{1}{f}$;

(iii) $3a - b = 3$, $2b - a = 1$;

(iv) $2 = \dfrac{1}{x} + \dfrac{3}{x^2}$.

5. The pressure of a gas at constant volume is proportional to its absolute (kelvin) temperature.

If the pressure of air in a tyre is 2.4×10^5 N/m^2 at 7°C (280 K), find (i) the temperature when the pressure increases to 2.6×10^5 N/m^2, (ii) the pressure when the temperature is 17°C.

6. (i) The shadow of a tree is 4.0 m long when the angle between the Sun's rays and the top X of the tree makes an angle of 60° with the horizontal (Figure 14.3(a)). Calculate the height h of the tree.

What is the length of the shadow when the angle shown in Figure 14.3(a) becomes 30°?

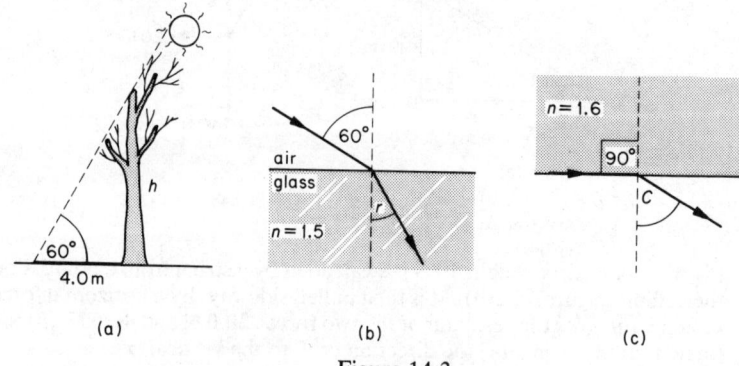

Figure 14.3.

(ii) Figure 14.3(b) shows a ray incident at 60° in air on an air–glass boundary. Find the angle of refraction r if the refractive index n of the glass is 1.5 ($\sin i / \sin r = n$). (iii) Figure 14.3(c) shows a ray in glass emerging in air at the critical angle C. Calculate C if the refractive index n *from air to glass* is 1.6 and $\sin C = 1/n$.

7. In Figure 14.4, $V = I(R + G)$. (i) Find V if $I = 0.005$, $R = 10$, $G = 990$. (ii) Calculate R if $V = 3$, $I = 0.01$, $G = 5$.

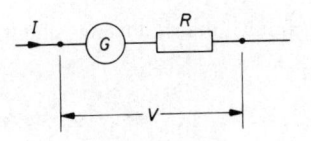

Figure 14.4.

8. The table below shows values of power P in watts (W) in a resistor for various values of current I in amperes (A).

P (W)	1.0	4.0	9.0	16.0	25.0	36.0
I (A)	0.5	1.0	1.5	2.0	2.5	3.0

(i) Plot P against I. (ii) From your graph, is P proportional to I? (iii) Calculate values of I^2 from each value of I, enter the results in the table, and plot P against I^2. From your graph, is $P \propto I^2$? (iv) If $P = RI^2$, find a value for R from your graph in (iii).

PAPER 5

1. In making an alloy, the following are used: 40% of a mass 1.5 kg of element X, 75% of a mass 2.8 kg of element Y, and 15% of a mass 3.2 kg of element Z.
   Calculate, to 2 s.f., the total mass of the alloy.

2. Calculate by logs, to 2 s.f.,

   (i) $\dfrac{0.88 \times 12.4}{121.2 \times 0.56^2}$,   (ii) $2 \times 3.142 \times \sqrt{\dfrac{0.56}{9.8}}$.

3. (i) A nail is struck by a force of 20.0 N at an angle of 30° to the vertical. Find the force driving the nail (a) downwards, (b) sideways.
   (ii) Water has a refractive index $n = 4/3$. If $n = \sin i/\sin r$ (Figure 14.5), calculate i when $r = 30°$.
   If $i = 90°$, calculate r.

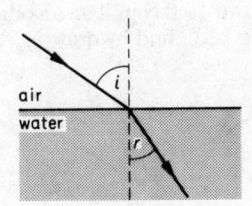

Figure 14.5.

170 BASIC MATHEMATICS FOR SCIENCE

4. (i) In an electric circuit, the following simultaneous equations are obtained:

$$\frac{E}{10+r} = \frac{2}{3}$$

and

$$\frac{E}{15+r} = \frac{1}{2}$$

Calculate E and r.

(ii) A lens equation between object distance u, image distance v and the lens focal length f is

$$\frac{1}{v} + \frac{1}{u} = \frac{1}{f}$$

Find (a) v if $u = +12$ cm, $f = -10$ cm, (b) f if $u = +20$ cm and $v = +30$ cm.

5. (i) In electrolysis, the quantity of charge or electricity Q in coulombs (C) passing through a solution is given by $Q = It$, where I is the current in amperes (A) and t is the time in seconds (s). Find the current flowing if a quantity 0.02 F passes in 500 s, where 1 F = 96 500 C.

(ii) A charge of 0.02 F is passed through copper(II) sulphate solution and dilute sulphuric acid. Calculate the mass of copper and the volume of hydrogen at s.t.p. produced. (Relative atomic mass: copper Cu = 63.5, hydrogen H = 1.)

6. The heat produced in a wire of given resistance for the same time is proportional to the square of the current flowing.

If the heat produced in these circumstances by a current of 0.5 A is 1500 J, calculate (i) the heat produced by a current of 1.0 A, (ii) the current which produces 24 000 J of heat.

7. A stationary ball of mass 0.15 kg is hit vertically upwards by a force F which is in contact with the ball for a time $t = 0.05$ s. The ball then moves with an initial speed $v = 30$ m/s.

Calculate

(i) the average acceleration a of the ball in m/s² if $v = at$,
(ii) the average force F in N (newton) if $F = ma$,
(iii) the gain in kinetic energy E of the ball if $E = \frac{1}{2}mv^2$,
(iv) the height h in m (metre) reached by the ball if $E = mgh$ and g has the numerical value 10.

8. (i) Two forces of 8.0 N and 10.0 N pull on an object O (Figure 14.6). If the angle between the forces is 120°, find by drawing their vector sum.

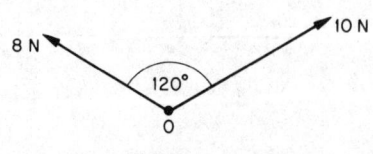

Figure 14.6.

(ii) The table below shows values of the actual depth, t, of a column of water in a cylinder and the measured apparent depth, a, which is less than the actual depth:

t/mm	150	200	240	300	350
a/mm	110	150	180	225	260

Plot a graph of t against a. (a) What is the apparent depth when $t = 275$ mm? (b) If the refractive index n of water $= t/a$, find a value for n from the slope of the graph.

Answers

Exercise 1 (p. 4)
1. $\frac{11}{20}$
2. $1\frac{23}{24}$
3. $1\frac{13}{20}$
4. $1\frac{19}{60}$
5. $2\frac{5}{18}$
6. $\frac{1}{20}$
7. $\frac{3}{4}$ kg
8. $\frac{1}{20}$
9. $1\frac{1}{10}$ kg

Exercise 2 (p. 6)
1. $\frac{2}{35}$
2. $\frac{25}{36}$
3. $4\frac{2}{3}$
4. $13\frac{1}{3}$
5. $15\frac{3}{7}$
6. 9
7. $33\frac{3}{4}$ km
8. $10\frac{1}{2}$ l
9. $\frac{3}{5}, \frac{9}{16}, \frac{27}{80}$ kg
10. $1\frac{1}{8}$ N m
11. (i) $3\frac{3}{20}$, $3\frac{3}{4}$ m, (ii) $11\frac{13}{16}$ m^2
12. $3\frac{3}{8}$ kg

Exercise 3 (p. 8)
1. $\frac{5}{6}$
2. $\frac{27}{64}$
3. $\frac{5}{12}$
4. $\frac{18}{31}$
5. $\frac{8}{11}$
6. $\frac{5}{6}$
7. $\frac{2}{3}$
8. $\frac{5}{6}$
9. (i) 60 km, (ii) 50 m.p.h., (iii) 80 km/h
10. 32
11. (i) $8\frac{8}{9}$ gal, (ii) 9 gal
12. 16

Exercise 4 (p. 11)
1. $\frac{3}{5}$
2. $\frac{1}{25}$
3. $\frac{3}{500}$
4. $9\frac{1}{2}$
5. $\frac{1}{200}$
6. $\frac{1}{250}$
7. $1\frac{2}{25}$
8. $2\frac{1}{5}$
9. $\frac{1}{1250}$
10. $5\frac{23}{100}$

Exercise 5 (p. 14)
1. 11.0
2. 13.4
3. 3.2
4. 3.7
5. 15.3
6. 2.5
7. 21
8. 1.03
9. 11.0
10. 1.24
11. £9.88
12. 25.41 kg
13. 10.17 m

ANSWERS

Exercise 6 (p. 17)
1. 0.1
2. 0.029
3. 57
4. 506
5. 1.8
6. 1.2
7. 0.003
8. 0.00 012
9. 0.61
10. 0.0012
11. 89.5
12. 48.4
13. 0.17 kg
14. 46.2 m^3
15. 12.6 l, 151 m^2
16. 74.1 g
17. 34.9 km
18. 4.6 m^2
19. (i) 45 km/h, (ii) 75 km/h
20. (i) 0.87, 0.87, 2.31 m, (ii) 1.7 m^3

Exercise 7 (p. 19)
1. 0.4
2. 0.07
3. 20
4. 30
5. 0.005
6. 1
7. 220
8. 0.24 cm
9. 11.4 g/cm^3
10. 520 J/kg K
11. 40
12. 0.004
13. 35
14. 0.25

Exercise 8 (p. 22)
1. 35 cm, 30 cm
2. 2.8 kg, 1.2 kg
3. (i) 1.28 kg, 0.72 kg, (ii) 180 g
4. 1.0 kg, 0.6 kg, 0.2 kg
5. £0.60, £0.40, £0.20
6. 84 g, 56 g, 280 g
7. 420 g, 240 g, 20 g

Exercise 9 (p. 24)
1. (i) $\frac{17}{20}, \frac{7}{10}, \frac{11}{20}, \frac{3}{8}, \frac{1}{20}$, (ii) 40, 78, $83\frac{1}{3}$, $52\frac{1}{2}$%
2. 350 million km^2
3. 80%
4. 105 W, 45 W
5. (i) 5000 J, (ii) 2000 J
6. (i) 24 000 J, (ii) 62.5%

Exercise 10 (p. 26)
1. (i) 42.9, (ii) 24.7, (iii) 52.9, (iv) 32.7, (v) 36.1%
2. YO, YO$_2$
3. XCl, XCl$_2$
4. XO, XO$_2$

Exercise 11 (p. 28)
1. FeCl$_3$
2. CH$_3$; C$_2$H$_6$
3. CuSO$_4$.5H$_2$O
4. FeCl$_2$.4H$_2$O
5. C$_2$H$_4$Br$_2$

Exercise 12 (p. 30)
1. 10^4
2. $6r^5$
3. $85y^8$
4. $2x^2$
5. $\frac{3}{x^3}$
6. $5a^{12}$
7. 108
8. 72
9. a^2

10. $\dfrac{y^4}{z^6}$
11. (i) 10^6, (ii) 24×10^7, (iii) 3×10^8, (iv) 3×10^3, (v) 2×10^3
12. (i) A, (ii) B, (iii) F
13. (i) 3.3×10^5 (ii) $86:1$
14. 3×10^8 m/s
15. 1.3 s
16. 0.15 N/mm^2
17. 12×10^{26}
18. 3×10^{19}/cm^3

Exercise 13 (p. 36)
1. (i) -3, (ii) 2, (iii) -6, (iv) 2
2. (i) 1, (ii) 3, (iii) -3, (iv) 2
3. (i) $-24°$C, (ii) $+2°$C
4. $\dfrac{1}{50}$
5. $\dfrac{2}{5}$
6. $\dfrac{3}{1000}$
7. $\dfrac{1}{4a^6}$
8. 20
9. 3×10^{-6}
10. 15
11. (i) 6×10^{-3} m, (ii) 5×10^{-2} m, (iii) 3×10^5 m, (iv) 5×10^{-7} m
12. (i) 5×10^{-6} m^2, (ii) 6×10^{-4} m^2, (iii) 2×10^{-5} m^2, (iv) 10^7 m^2
13. 3×10^{-4} m, 2×10^{-7} m
14. 5×10^{-7} m
15. 5×10^{14} Hz
16. 3×10^{-5} s, 6×10^4 m
17. (i) 8.8×10^{-3} kg, (ii) 0.66 kg
18. 96×10^{-4} m^2
19. 5.12×10^{-7} m^2

Exercise 14 (p. 39)
1. 2
2. 3
3. 9
4. 1000
5. 100
6. $2a$
7. $3y^2$
8. $8x^3$
9. $\dfrac{1}{10}$
10. $\dfrac{1}{1000}$
11. $\dfrac{1}{4}$
12. $\dfrac{1}{8}$

Exercise 15 (p. 42)
1. (i) 62.8 mm, (ii) 314 mm^2, (iii) 628 mm^3
2. (i) $2:1$, (ii) $4:1$, (iii) $8:1$
3. 0.28 mm^2
4. 50 mm^2, 33 mm^2
5. (i) $1:4$, (ii) $1:8$
6. (i) 1005 cm^2, (ii) 1005 cm^3
7. (i) $2:1$, (ii) $8:1$
8. 27, 4400 cm^3
9. $10:1$

Exercise 16 (p. 45)
1. 1.8042
2. 0.1712
3. 2.8726
4. 1.4995
5. 4.9197
6. 451.9
7. 3.142
8. 74.91
9. 1194
10. 4.766×10^6

Exercise 17 (p. 46)
1. 139.3
2. 574.2
3. 5.5
4. 24.1
5. 21.3

Exercise 18 (p. 48)
1. 0.021
2. 25

3. 0.0067
4. 0.070
5. 4.6

Exercise 19 (p. 51)
1. 0.000 64
2. 0.84
3. 0.97
4. 0.0032
5. 1.7

Exercise 20 (p. 57)
1. $a = 5$
2. $E = 3\frac{2}{3}$
3. $y = 2$
4. $I = 1$
5. $r = 3$
6. $x = 9$
7. $x = 10$
8. $a = 0$
9. $I = 2$
10. $x = 5$

Exercise 21 (p. 60)
1. $x = 2\frac{1}{3}$
2. $v = 3$
3. $R = 2$
4. $s = \frac{6}{7}$
5. $y = 11\frac{1}{2}$
6. $a = 11$
7. $R = 8$
8. $u = \frac{9}{10}$
9. $u = \frac{4}{9}$
10. $x = 13\frac{1}{2}$

Exercise 22 (p. 62)
1. (i) 45 m/s, (ii) 300 m
2. (i) 40 m/s, (ii) 200 m
3. 16 m/s
4. 20 m/s
5. (i) 2 s, (ii) 20 m

Exercise 23 (p. 65)
1. 1 N, 2 N
2. 3000 N, 10 000 N
3. 240 kg m/s, 80 N
4. (i) 1 kg m/s, (ii) 1 N
5. (i) 12 000 kg m/s, (ii) 5 m/s^2, (iii) 6000 N
6. 50 N
7. 1800 N

Exercise 24 (p. 67)
1. 200 000 J
2. (i) 500 J, (ii) 500 J
3. 80 J
4. 60 000 J
5. 8000 J, 100 000 J
6. 12 m
7. 15.5 m/s
8. 20 m/s

Exercise 25 (p. 70)
1. 30 V
2. 0.4 A
3. 1.4 Ω
4. 0.5 A
5. 1.8 V
6. 450 C
7. 2.5 A
8. 965 s
9. 0.025 F
10. 3.3×10^{-4} g/C

Exercise 26 (p. 72)
1. (i) 7 Ω, (ii) 1.7 Ω
2. 4.5 Ω
3. (i) 1.2 A, (ii) 5 A
4. 125 Ω
5. 100 Ω
6. 3.2 Ω

Exercise 27 (p. 73)
1. 31 680 J
2. 750 J

3. 1440 J
4. 60 W
5. 100 Ω, 1 W
6. 25 W

Exercise 28 (p. 77)
1. 4400 J, 880 J/K
2. 114 750 J
3. (i) 12 000 J, (ii) 1500 J/K, (iii) 750 J/kg K
4. 4 kg, 85 000 J
5. 500 s
6. 13 200 J, 0.04 kg
7. (i) 22 000 J, (ii) 23 680 J, (iii) 474 J/kg K
8. (i) 3300 J, (ii) 3510 J, (iii) 9 K (°C)

Exercise 29 (p. 83)
1. 37.5 cm
2. 30 cm, virtual image
3. 8 cm
4. 9.6 cm, virtual
5. 10 cm
6. 15 cm
7. 8 cm
8. 6 cm from lens
9. 6.9 cm
10. 20 cm

Exercise 30 (p. 89)
1. $(v-u)/t$
2. $gT^2/4\pi^2$
3. $\sqrt{A/\pi}$
4. $\sqrt{q^2+r^2}$
5. F/m
6. $p/\rho g$
7. $(mu-p)/M$
8. $\sqrt{2s/g}$
9. $\sqrt[3]{3V/4\pi}$
10. $2(s-ut)/t^2$
11. -10 m/s^2

12. 0.25 m
13. 1.5 m/s^2
14. 20 m/s
15. 8 m
16. 2.4 m/s
17. (i) 6 m/s, (ii) 7 m/s
18. 10 N
19. 15 m
20. 6 m

Exercise 31 (p. 94)
1. RA/l
2. V/I
3. $(E-Ir)/I$
4. $\sqrt{P/R}$
5. RA/ρ
6. $\sqrt{PR}$
7. $(W-w)/IV$
8. $(2E-IR)/2I$
9. $\pi R r^2/\rho$
10. $R(E-V)/V$
11. 3 Ω
12. 0.5 A
13. 12 V, 3 A
14. 2 Ω
15. 9 V
16. 62.5 Ω
17. 4 A
18. 240 V
19. 1.2 m
20. 2×10^{-6} Ω m

Exercise 32 (p. 98)
1. (i) 150 000 J, (ii) 6.7°C
2. 1367 J/kg K
3. 5.6 kg
4. $\frac{1}{3}$ kg
5. 1247 J/kg K
6. 20 150 J

Exercise 33 (p. 103)
1. $x=4, y=1$
2. $x=1, y=2$

ANSWERS

3. $x = 3, y = 1$
4. $x = 2, y = 3$
5. $x = 5, y = 3$
6. $x = 2, y = 2$
7. (i) $E = 12, r = 2$, (ii) $E = 1.5$, $r = 4$, (iii) $E = 2, r = 1$
8. 5.1 cm
9. 10 cm
10. 3150 J/min, 420 J/K

Exercise 34 (p. 106)
1. 2 or -1
2. 1 or $\frac{1}{2}$
3. 2 or $-\frac{1}{3}$
4. 1.2 or -1.7
5. 1.8 or -1.1
6. 2 s, 3 s
7. 0.8 s, 7.2 s
8. 5 cm
9. 3.4 A

Exercise 35 (p. 110)
1. (iii) (a) Aug, (b) Apr/May
2. (iii) (a) 870, (b) 22
3. (i) 35°C, (ii) 280 g/l, (iii) 22°C, (iv) *decreases* with temp. rise, (v) 15 g/l per °C
4. (i) S, (ii) -32°C, (iii) -24°C, 12%, 24°C, 50%, (iv) S
5. (i) 3.1 s, (ii) 31 m, (iii) 15 m/s, (iv) 35 m/s
6. (ii) 5.0, 5 Ω, (iii) 0.9 Ω, (iv) 2.4 Ω
7. (b) 45°C, (c) 53 units/min
8. (ii) 130/min, (iv) 21/min, (vi) 1 min
10. (ii) (a) about 28%, (b) about 1.3 years
11. (ii) (a) about 40 days, (b) about 165 days
12. (ii) (a) about 8.5, (b) about 14.5 mg/cm^2

Exercise 36 (p. 118)
1. (i) 4, (ii) no
2. (i) 2, (ii) yes
3. (i) -3, (ii) no
4. (i) -4, (ii) yes
5. (i) 1.5, (ii) no

Exercise 37 (p. 123)
1. (ii) 10 Ω
2. (ii) 4 N/mm
3. (ii) $F \propto a$ (iii) 2 kg
4. 10 m/s^2
5. $Q \propto I^2$
6. $R \propto 1/A$
7. $E \propto 1/r^2$
8. $P \propto V^2$

Exercise 38 (p. 129)
1. (i) 0.96 mm, (i) 31 N
2. (i) 4 : 1, (ii) 8 : 1
3. (i) 4.0 Ω, (ii) 0.9 m
4. (i) 1.1×10^5 N/m^2, (ii) 328 K
5. (i) 54.2 cm^3, (ii) 111 cm^3
6. (i) 72 000 J, (ii) 1.0 A
7. (i) 1600 Hz, (ii) 0.8 N
8. (i) 20 m/s, (ii) 80 J
9. (i) 0.56 m, (ii) 3.2 s
10. (i) 0.7 A, (ii) 1.0 V

Exercise 39 (p. 135)
1. (i) 500 Hz, (ii) 15.0 cm
2. (i) 2.5 Ω, (ii) 0.1 mm
3. (i) 120 cm^3, (ii) 1.8×10^5 N/m^2
4. (i) 6.4 N, (ii) 12 800 km
5. (i) 52 W, (ii) 390 Ω
6. (i) $R = kl/A$, (ii) 24 Ω, (iii) 2.25 m
7. (i) $P \propto V^2/l$, (ii) 0.6 m, (iii) 15 V
8. (i) 0.3 A, (ii) 0.9 Ω

Exercise 40 (p. 140)
1. 0.1 g, (ii) 17.75 g
2. (i) 40 cm^3, (ii) 448 cm^3
3. 2 mol
4. (i) 0.2 g, (ii) 6.35 g, (iii) 2.24 dm
5. 7
6. (i) 0.8 g, (ii) 560 cm^3, (iii) 4.8 g

Exercise 41 (p. 143)
1. (i) H$^+$, Ag$^+$, Cu^{2+}, Cl$^-$, Al^{3+}, (ii) +e, +e, +2e, −e, +3e, (iii) $\frac{1}{2}$, 1, $\frac{1}{2}$, 1, $\frac{1}{3}$ mol
2. (iii) 11.2 dm^3
3. (i) 10.8 g, (ii) 0.1 g, (iii) 1.12 dm^3
4. (i) 1.5 F, (ii) 0.1 F, (iii) 1 F
5. 63.6 g, 5.4 g
6. 0.01 F
7. (i) 900 C, (ii) 0.3 g

Exercise 42 (p. 152)
1. sin A = 4/5, sin B = 3/5
2. sin A = 12/13, sin B = 5/13
3. 0.68 m, 1.37 m
4. 30 m, 10 m
5. 28°
6. 42°
7. 36°
8. 1.4

Exercise 43 (p. 158)
1. (i) 4/5, (ii) 3/5
2. (i) 5/13, (ii) 12/13, (iii) 12/13, (iv) 5/13, (v) 1
3. 71 N, 87 N
4. 87 N, 100 N
5. 115 N
6. (i) $F/2$, (ii) 87 N, (iii) 173 N
7. 69 N, 40 N; 51°

Exercise 44 (p. 163)
1. (i) 4/3, (ii) 3/4
2. 3.6
3. (i) 32°, (ii) 9.4
4. (i) 37°, (ii) 2.5 km
5. (i) 56°, (ii) 36 N; 36 N
6. (i) 7°, (ii) 806 km/h
7. (i) 50.3 N, (ii) 78.3 N
8. (i) 27°, (ii) 1.1 m/s
9. 30°

REVISION PAPERS

Paper 1 (p. 165)
1. (i) $1\frac{5}{8}$, (ii) $3\frac{7}{10}$
2. (i) 0.13, (ii) 0.4
3. (i) $x = 3$, (ii) $x = 1\frac{1}{7}$
4. (i) 40, (ii) 2
5. 40.9 cm^3
6. (i) 2.45 s, (ii) 3.1 m
7. (i) 4.3 cm, 2.5 cm, (ii) (a) 50 N, (b) 5 m/s^2
8. (ii) $V \propto I$, (iii) $V = 2.0$ V, (iv) 1.7 Ω

Paper 2 (p. 166)
1. (i) $2\frac{4}{9}$, (ii) $1\frac{1}{2}$
2. (i) 0.047, (ii) 0.53
3. (i) 8.4, (ii) 94.4
4. (i) $y = 2\frac{1}{2}$, (ii) $v = -60$
5. (i) 10^6, (ii) 2×10^{-2}
6. (i) $(v-u)/t$, (ii) 12
7. (i) 9 Hz, (ii) 0.45 m
8. $y \propto 1/x$

Paper 3 (p. 166)
1. (i) $2\frac{4}{5}$, (ii) 5
2. (i) 0.15, (ii) 24
3. (i) 0.73, (ii) 0.13
4. (i) $a = 2\frac{1}{2}$, (ii) $u = 48$
5. (i) 1.2 kg m/s, (ii) 8.4 J

6. (i) $\frac{1}{3}$ A, (ii) 2 Ω
7. (i) 32 Ω, (ii) 1.4 mm
8. (i) $x = 6.9$ cm, $y = 8.0$ cm, (ii) (a) 44.7 N, (b) 44.7 N, (c) 27°

Paper 4 (p. 167)
1. (i) $4\frac{3}{40}$, (ii) $4\frac{11}{16}$
2. (i) 0.09, (ii) 9.6, (iii) 4.1
3. (i) 54%, (ii) $NaNO_3$
4. (i) $R = 1$, (ii) $f = 4.8$, (iii) $a = 1.4$, $b = 1.2$, (iv) $x = 1\frac{1}{2}$ or -1
5. (i) 303 K, (ii) 2.5×10^5 N/m²
6. (i) 6.9 m, 12.0 m, (ii) 35°, (iii) 39°
7. (i) $V = 5$, (ii) $R = 295$
8. (ii) no, (iii) yes, (iv) 4

Paper 5 (p. 169)
1. 3.18 kg
2. (i) 0.29, (ii) 1.5
3. (i) (a) 17.3 N, (b) 10.0 N, (ii) 42°, 49°
4. (i) $E = 10$, $r = 5$, (ii) (a) −5.5 cm, (b) 12 cm
5. (i) 3.86 A, (ii) 0.64 g, 0.224 dm³
6. (i) 6000 J, (ii) 2.0 A
7. (i) 600 m/s², (ii) 90 N, (iii) 67.5 J, (iv) 45 m
8. (i) 9.2 N, (ii) (a) 205 mm, (b) 1.33